SpringerBriefs in Space Life Sciences

Series Editors

Günter Ruyters
Space Administration, German Aerospace Center (DLR)
Bonn, Germany

Markus Braun
Space Administration, German Aerospace Center (DLR)
Bonn, Germany

The extraordinary conditions of space, especially microgravity, are utilized for research in various disciplines of space life sciences. This research that should unravel – above all – the role of gravity for the origin, evolution, and future of life as well as for the development and orientation of organisms up to humans, has only become possible with the advent of (human) spaceflight some 50 years ago. Today, the focus in space life sciences is 1) on the acquisition of knowledge that leads to answers to fundamental scientific questions in gravitational and astrobiology, human physiology and operational medicine as well as 2) on generating applications based upon the results of space experiments and new developments e.g. in non-invasive medical diagnostics for the benefit of humans on Earth. The idea behind this series is to reach not only space experts, but also and above all scientists from various biological, biotechnological and medical fields, who can make use of the results found in space for their own research. SpringerBriefs in Space Life Sciences addresses professors, students and undergraduates in biology, biotechnology and human physiology, medical doctors, and laymen interested in space research. The Series is initiated and supervised by Dr. Günter Ruyters and Dr. Markus Braun from the German Aerospace Center (DLR). Since the German Space Life Sciences Program celebrated its 40th anniversary in 2012, it seemed an appropriate time to start summarizing – with the help of scientific experts from the various areas - the achievements of the program from the point of view of the German Aerospace Center (DLR) especially in its role as German Space Administration that defines and implements the space activities on behalf of the German government.

More information about this series at http://www.springer.com/series/11849

Günter Ruyters • Markus Braun •
Katrin Maria Stang

Breakthroughs in Space Life Science Research

From Apollo 16 to the ISS

 Springer

Günter Ruyters (Retired)
Rheinbreitbach, Germany

Markus Braun
German Space Agency, Research
and Exploration
German Aerospace Center (DLR)
Bonn, Germany

Katrin Maria Stang
German Space Agency, Research
and Exploration
German Aerospace Center (DLR)
Bonn, Germany

ISSN 2196-5560   ISSN 2196-5579   (electronic)
SpringerBriefs in Space Life Sciences
ISBN 978-3-030-74021-4   ISBN 978-3-030-74022-1   (eBook)
https://doi.org/10.1007/978-3-030-74022-1

This Springer imprint is published by the registered company Springer Nature Switzerland AG.
The registered company address is: Gewerbestrasse 11, 6330 Cham, Switzerland

Foreword

Twelve books have been published since 2015 in the series "SpringerBriefs in Space Life Sciences." Internationally renowned scientists have covered basically all areas of space life sciences research focusing on the effects of changing gravity conditions from single cells up to human beings, including also radiation effects, astrobiology, and biotechnology. In this 13th and final book of the series, the authors present their personal view on the scientific accomplishments—be it important incremental progress or greater breakthroughs—as well as on the technological achievements from roughly 40 years of space life sciences especially in Europe and in Germany.

In Chap. 1, the scene is set by describing the extraordinary conditions of space and by giving a historical background on the establishment of space medicine and space life sciences in general with new fascinating research aspects. The introductory chapter is complemented with organizational and programmatic issues of the implementation process, especially in Europe after World War II.

Chapter 2 provides a detailed overview of the flight opportunities available for space life science research in Europe ranging from the Drop Tower in Bremen, via parabolic flights with airplanes, sounding rockets such as TEXUS, recoverable satellites such as the Russian BION and FOTON as well as the Chinese Shenzhou, the Shuttle/Spacelab program, the Russian space station MIR, and finally to the International Space Station ISS. Information on simulation devices for biological research and on opportunities for performing human physiology experiments such as bedrest, isolation, and confinement studies that prepare or complement space research concludes this chapter.

Chapters 3 and 4 represent the central part of the book emphasizing success stories in space life science research and in technology development. In Chap. 3, incremental progress in knowledge as well as greater scientific breakthroughs are described. Key findings in gravitational biology such as the elucidation of the signal transduction chain for gravitaxis and gravitropism as well as in radiation biology, astrobiology, microbial challenges, and protein crystallization are covered at first. In the second part of Chap. 3, key findings and breakthroughs in human physiology and space medicine are presented. Important results on microgravity-related changes of the cardiovascular, the neurovestibular, the musculoskeletal and the immune systems are summarized elaborating also on the mechanisms behind. One of the highlights is certainly the "salt story"—a completely new integrative view on the context of salt and nutrition with the immune, the cardiovascular, the bone and muscle systems leading to new approaches and strategies for fighting certain diseases such as hypertension and osteoporosis. A brief look at human health and performance issues concludes the chapter emphasizing the similarity between spaceflight-induced changes in astronauts and aging processes on Earth as nicely depicted by the slogan "Geriatrics meets Spaceflight." At the end, a concise table provides an impressive summary of all the key findings described.

In Chap. 4, the authors present further success stories—focusing here on innovative developments for biomedical diagnostics and preventive healthcare. Quite a number of new devices and technologies as well as countermeasures for fighting diseases and for rehabilitation have been developed not the least in the ESA and DLR Life Sciences programs. Many of them have found their way into the commercial market with application potential in various fields such as the clinic, sports, rehabilitation, and disease monitoring to name a few.

While the focus certainly is on European success stories, and here especially on those from the German program that has played a leading role over decades in Europe, the wealth of international references provided pays tribute to further worldwide accomplishments.

The book closes in Chap. 5 with a brief outlook on future challenges and opportunities in the upcoming exploration era. Ambitious research and development activities on thematically relevant topics such as bioregenerative life support systems as well as human health and performance are required in space life sciences to enable and support the efforts of mankind to leave Earth behind and reach out for Moon, Mars, and other distant destinations. At the same time, these exploration missions and future stations on the surface of Moon or Mars will also offer exciting new opportunities. Future achievements will again be of benefit for people on Earth as is presented in this volume demonstrating the accomplishments of space life science and technology of the past few decades.

Rheinbreitbach, Germany   Günter Ruyters
Bonn, Germany   Markus Braun
March 2021

Preface to the Series

The extraordinary conditions in space, especially microgravity, are utilized today not only for research in the physical and materials sciences—they especially provide a unique tool for research in various areas of the life sciences. The major goal of this research is to uncover the role of gravity with regard to the origin, evolution, and future of life, and to the development and orientation of organisms from single cells and protists up to humans. This research only became possible with the advent of manned spaceflight some 50 years ago. With the first experiment having been conducted onboard Apollo 16, the German Space Life Sciences Program celebrated its 40th anniversary in 2012—a fitting occasion for Springer and the DLR (German Aerospace Center) to take stock of the space life sciences achievements made so far.

The DLR is the Federal Republic of Germany's National Aeronautics and Space Research Center. Its extensive research and development activities in aeronautics, space, energy, transport, and security are integrated into national and international cooperative ventures. In addition to its own research, as Germany's space agency the DLR has been charged by the federal government with the task of planning and implementing the German space program. Within the current space program, approved by the German government in November 2010, the overall goal for the life sciences section is to gain scientific knowledge and to reveal new application potentials by means of research under space conditions, especially by utilizing the microgravity environment of the International Space Station (ISS).

With regard to the program's implementation, the DLR Space Administration provides the infrastructure and flight opportunities required, contracts the German space industry for the development of innovative research facilities, and provides the necessary research funding for the scientific teams at universities and other research institutes. While the so-called small flight opportunities like the drop tower in Bremen, sounding rockets, and parabolic airplane flights are made available within the national program, research on the ISS is implemented in the framework of Germany's participation in the ESA Microgravity Program or through bilateral cooperations with other space agencies. Free flyers such as BION or FOTON satellites are used in cooperation with Russia. The recently started utilization of

Chinese spacecrafts like Shenzhou has further expanded Germany's spectrum of flight opportunities, and discussions about future cooperation on the planned Chinese Space Station are currently underway.

From the very beginning in the 1970s, Germany has been the driving force for human spaceflight as well as for related research in the life and physical sciences in Europe. It was Germany that initiated the development of Spacelab as the European contribution to the American Space Shuttle System, complemented by setting up a sound national program. And today Germany continues to be the major European contributor to the ESA programs for the ISS and its scientific utilization.

For our series, we have approached leading scientists first and foremost in Germany, but also—since science and research are international and cooperative endeavors—in other countries to provide us with their views and their summaries of the accomplishments in the various fields of space life sciences research. By presenting the current SpringerBriefs on muscle and bone physiology, we start the series with an area that is currently attracting much attention—due in no small part to health problems such as muscle atrophy and osteoporosis in our modern aging society. Overall, it is interesting to note that the psycho-physiological changes that astronauts experience during their spaceflights closely resemble those of aging people on Earth but progress at a much faster rate. Circulatory and vestibular disorders set in immediately, muscles and bones degenerate within weeks or months, and even the immune system is impaired. Thus, the aging process as well as certain diseases can be studied at an accelerated pace, yielding valuable insights for the benefit of people on Earth as well. Luckily for the astronauts: these problems slowly disappear after their return to Earth, so that their recovery processes can also be investigated, yielding additional valuable information.

Books on nutrition and metabolism, on the immune system, on vestibular and neuroscience, on the cardiovascular and respiratory system, and on psycho-physiological human performance will follow. This separation of human physiology and space medicine into the various research areas follows a classical division. It will certainly become evident, however, that space medicine research pursues a highly integrative approach, offering an example that should also be followed in terrestrial research. The series will eventually be rounded out by books on gravitational and radiation biology.

We are convinced that this series, starting with its first book on muscle and bone physiology in space, will find interested readers and will contribute to the goal of convincing the general public that research in space, especially in the life sciences, has been and will continue to be of concrete benefit to people on Earth.

DLR Space Administration in Bonn-Oberkassel (DLR)

The International Space Station (ISS). (Photo taken by an astronaut from the space shuttle Discovery, March 7, 2011 (NASA))

Extravehicular activity (EVA) of the German ESA astronaut Hans Schlegel working on the European Columbus lab of ISS, February 13, 2008 (NASA)

Bonn, Germany   Günter Ruyters
July 2014   Markus Braun

Acknowledgments

Life Sciences research in space is not just a great field of research but also a passion—a passion the authors of this issue share with the many authors of the 12 previous books of the series SpringerBriefs in Space Life Sciences. We are grateful to all of them for many exciting discussions in the various phases of the genesis of the manuscripts; a summary of the accomplishments from our perspective is presented in this 13th and final volume.

Space projects in general, but even more so Space Life Sciences projects might start with clever ideas of individual scientists, but the perfect implementation and ultimately the successful exploitation and publication of results is always based on the precisely orchestrated interaction of teams from many different disciplines. We thank our colleagues and friends from many space agencies—especially those assembled in the International Space Life Sciences Working Group (ASI, CNES, CSA, DLR, ESA, JAXA, Roscosmos and NASA) and from the DLR Life Sciences and Microgravity program. We also wish to express our sincere thanks to the Russian colleagues at IBMP Moscow and to our counterparts in Chinese space organizations. They all fostered international, bi-, or multilateral cooperation that is absolutely mandatory for success in these endeavors. Numerous engineers and technicians, astronauts and cosmonauts, as well as scientists and university students contributed to the space projects by the compilation of concepts, the development and testing of experiment hardware and not to forget by the vast amount of paperwork in all phases of ground-based research projects and spaceflight missions. Scientific success of such complex missions can only be achieved through great commitment of all those involved and the absolute will to deliver best performance—not to forget that of the humans flying into space. Of course, all this is only possible by a substantial financial investment. In this context, we acknowledge in particular the funding of projects provided by DLR Space Administration (German Aerospace Center) on behalf of the Bundesministerium für Wirtschaft und Energie (BMWi).

Over the years and decades, colleagues and friends also made us familiar with the culture and the mode of living and working in their respective countries. Unforgettable experiences and events and long-lasting friendships are our personal priceless gains.

Contents

About the Authors

Günter Ruyters is the former Head of Germany's Space Life Sciences Program at the German Space Administration, German Aerospace Center DLR and served as Germany's delegate of the Human Spaceflight, Microgravity and Exploration Program Board of the European Space Agency ESA. He was also appointed professor at the Faculty of Biology, University of Bielefeld.

Markus Braun is Head of Germany's Space Life Sciences Program at the German Space Agency, German Aerospace Center (DLR), and serves as Germany's delegate of the Human Spaceflight, Microgravity and Exploration Program Board of the European Space Agency ESA. He is appointed faculty member at the University Bonn and is principal investigator and project manager of gravitational biology, physiology and bioregenerative life support projects conducted on various microgravity flight opportunities.

Katrin Maria Stang is Parabolic Flight Program Manager and Coordinator for the Human Research Program at the Department of Research and Exploration at the German Space Agency, German Aerospace Center (DLR). Her responsibilities include management of experiment hardware development for the International Space Station and coordination of the utilization of ground analogues for human physiology projects.

Chapter 1
Introduction: Space Life Sciences—Basic Research and Applications Under Extraordinary Conditions

Abstract Sixty years of human spaceflight after the maiden flights of Yuri Gagarin in 1961 and John Glenn in 1962, respectively, seem to be justification enough to look back and summarize the scientific results and technological achievements in space life sciences. In the introductory chapter, the scene is set by describing the extraordinary environmental conditions of space and their importance for living systems. Also, the historical background for establishing space life sciences especially after World War II with the foundation of ESA and national European space agencies and of their respective programs is provided. A short look on the activities of the former German Democratic Republic (GDR) closes the introduction.

Keywords DLR · ESA · History of space life sciences · Interkosmos · Space conditions

1.1 Rationale and Introductory Remarks

Sixty years of human spaceflight after the maiden flights of Yuri Gagarin in 1961 and John Glenn in 1962, respectively, seem to be justification enough to look back and summarize the scientific results and technological achievements in space life sciences. Moreover, space life sciences activities in Germany are close to celebrating their 50th anniversary after the first experiment in radiation biology onboard of Apollo 16 in the year 1972. On the other hand, looking into the future, space faring nations today are on the edge of entering a new exploration era. After the positive experiences with the ISS these activities will most probably be conducted in an international framework of cooperation and coordination.

To some extent, this transition period had also stimulated a few years ago our thoughts to set up—together with Springer—the SpringerBrief Series in Space Life Sciences (https://www.springer.com/series/11849). Since 2015, 12 volumes have been published so far. In addition to covering radiation-related and astrobiological topics as well as biotechnological aspects, nine books deal with the effects of changed gravity conditions, above all microgravity, on living systems—especially on the human body emphasizing astronaut physiology, health, and performance.

G. Ruyters et al., *Breakthroughs in Space Life Science Research*, SpringerBriefs in Space Life Sciences, https://doi.org/10.1007/978-3-030-74022-1_1

While these books present in detail a wealth of scientific and application-oriented results, the current volume tries to give a comprehensive summary with a focus on scientific breakthroughs and outstanding technological developments mainly in Europe for the benefit of people on Earth. But before doing so, the scene is set by describing the extraordinary conditions of space and their stimulating effect on life sciences research as well as by providing some clue on the historical background.

1.2   Extraordinary Conditions in Space Stimulating Life Science Research

The physical conditions in space are extreme and extraordinary. While spacecrafts with their life support systems have been developed to largely protect astronauts and other living beings as well as the technical infrastructure from the hazardous challenges of the space vacuum, high temperature changes, and the complex and dangerous radiation fields, the physical factor "gravity" is totally absent within a spacecraft circulating the Earth. It is this condition that stimulated life (and physical) sciences research over decades. Why is this so?

Gravity is ubiquitous on Earth. From simple physical or chemical to complex biological systems—it influences everything that happens on our Earth. Frequently, this is more than obvious: objects fall to the ground, water flows downhill, and gas bubbles rise in boiling water. In other natural and technical processes however gravity's influence is not immediately apparent so that its significance may only be uncovered through experiments in microgravity. Moreover, terrestrial gravity and life have been inseparably linked in the evolutionary process on our planet for around 4 billion years. Scientists wishing to learn about the part played by a given factor commonly modify its intensity or eliminate it altogether. This means in our case: we must get rid of gravity; we have to go into weightlessness.

In physical terms, an object is weightless when it is in free fall. A ball thrown into the air is similarly in a state of free fall, meaning that it is weightless. It is flying along a so-called ballistic trajectory. Generally, all states of weightlessness represent forms of free fall. This is not only true for capsules falling in a drop tower or for sounding rockets. Also, the orbit of satellites or of a space station around the Earth can be visualized as a "parabolic trajectory" that encircles the Earth. In orbit, momentum equalizes the gravitational pull of the Earth, which is still practically undiminished even at an altitude of 400 km. The result: all objects are weightless. Astronauts float freely in space.

However, free fall is an ideal condition that is almost never found in reality. All falling bodies are exposed to spurious accelerations of different intensity caused by, for instance, air drag and natural vibrations. Therefore, instead of weightlessness, the term "microgravity" has come to be used to describe extremely low gravity.

On Earth however gravity can be nullified only for a short time. Experiments can be conducted in drop towers such as in Bremen, Germany, or by using sounding rockets like TEXUS that provides microgravity for about 6 min. If longer periods of microgravity are needed, science must leave Earth behind and go to space, currently using research satellites like FOTON, BION, and Shenzhou, the International Space Station, or the so-called Heavenly Palace—the Chinese space station Tiangong. The only chance for a non-astronaut scientist to be personally present at experiments in microgravity is on parabolic flights using an airplane. To sum up—only (human) spaceflight provides the means and conditions to elaborate the role of gravity for life on Earth in detail (see Chap. 2 for further details on the respective flight opportunities).

1.3   Exploration of Basic Vital Functions and New Methods in Medical Diagnostics and Treatment

As already mentioned, life on Earth is subject to the influence of gravity, its significance for many vital functions can be explored only in microgravity. Scientists observe how organisms react and how biological processes change under these conditions. From their experiments in microgravity, researchers learn about the mechanisms by which living beings, from protozoa to humans, perceive and respond to gravity. Reaching beyond basic biology, these experiments in microgravity also permit conclusions as to the origin, spread, and development of life on our home planet.

Moreover, human life is basically an everlasting struggle against gravity, from a baby learning to walk to the infirmities of old and sick people. Therefore, medicine, too, benefits from research in microgravity: it provides new insights into the interplay between the various systems of the human body, e.g., between muscles and bones, the heart and the circulatory system, and the immune system. What happens to the astronauts onboard the ISS is changing our knowledge about the human body. Besides entering into the diagnosis and treatment of the sick, such knowledge also helps to preserve people's health and physical capacity in our mobility-dependent— and increasingly aging—society. Thus, research under space conditions has already spawned new training methods, new therapies, and new instruments for the benefit of people on Earth. Both these lines of research—the exploration of basic vital functions and the development of new methods for preventive health care, of diagnosis and treatment of illness in medicine—will also play a crucial role in preparing for future long-term missions to the Moon or to other destinations. We will elaborate these aspects in much more detail below. But—how did space life sciences come to origin?

1.4   How Things Began: Establishment of Space Life Sciences

In 1923, physicist Hermann Oberth presented his book "Rockets to Interplanetary Space" (Oberth 1923, English translation 2014). Independently of the work done by Tsiolkovsky in Russia and Goddard in the USA, he explained that a rocket would be a suitable vehicle for carrying man into space. Respective technological developments originated in the 1930s mainly in Germany, especially after Adolf Hitler had been appointed "Reichskanzler" in Germany end of January 1933. Already in 1934, Wernher von Braun, the father of the later Moon landing, succeeded in establishing a financially secure long-term rocket research program in the frame of Hitler's accelerated re-armament efforts. An ultramodern research and development center was built at Peenemünde on the Baltic Sea island of Usedom in 1936/1937 and was used to lay the foundations of this new technology between 1936 and 1945. Although the first chapter in the history of astronautics has certainly been the darkest so far, it was to leave its mark on the developments in the second half of the twentieth century (for more details see Reinke 2007).

At the end of World War II, the Allies—especially the Soviet Union and the United States and, to a lesser extent, also France and Great Britain—were engaged in a frantic race for the scientific knowledge of the Nazi state, especially regarding the German rocket program. Many of the Peenemünde technicians were transferred to the Soviet Union, which had a long tradition of rocket research dating back to Tsiolkovsky's time. A group of about 130 engineers led by van Braun and his coworkers were taken to the US at the end of 1945 in a top-secret operation codenamed "Overcast." Although the US had a 30-year tradition of rocket research that had begun with Goddard, no American space program existed by 1946. The transfer of German scientists and of the technology involved in their removal to the US soon proved extremely valuable for the American rocket program. From 1946 to 1957, many experts from Germany were accordingly offered normal contracts of employment with the American forces (Operation "Paperclip").

On October 4, 1957, the Soviet Union launched the first artificial satellite, "Sputnik." The successful launch of the Sputnik satellite shocked the USA out of its complacency regarding its technological superiority. As part of the rush to recover the initiative, NASA was promptly created in 1958. In May 1961, President Kennedy announced in his speech before Congress that it was an American national objective to land an American on the Moon before 1970: "I believe that this nation should commit itself to achieving the goal, before this decade is out, of landing a man on the moon and returning him safely to the Earth" (www.history.nasa.gov/moondec.html). To achieve this goal NASA's Apollo program was initiated. With the successful landing on the Moon in July 1969, the USA again demonstrated their superiority in high technology and obtained their lead in the space race that the Soviet Union had in possession due to the Sputnik flight and that of the first man, Yuri Gagarin, in an Earth orbit on April 12, 1961.

In these days, space life sciences or—more precisely—biomedical space research had already established itself in the USA and the Soviet Union. At that time however neither the Americans nor the Soviets were concerned with the overall biomedical or physiological effects of microgravity from a scientific point of view. Their interest focussed on securing the ability of humans to survive in space and preserving their health and physical performance. The first two flights, undertaken by Yuri Gagarin in April 1961 and John Glenn in February 1962, had demonstrated that humans are able to survive a sojourn in space. Yet subsequent space missions revealed that astronauts are likely to be troubled by a number of health problems, especially space sickness. For this reason, medical projects that addressed these problems enjoyed top priority. As preparation for human spaceflight missions, also several animal experiments were performed with dogs, mice, and monkeys by both the USA and the Soviet Union—some readers may still remember the dog Laika on the Sputnik mission of November 1957. Other living beings were brought into space very soon after that. For instance, plants got to take space trips onboard Sputnik 4 and Discovery 17 as early as 1960.

The interest in space life sciences research, especially in space medicine however dates back much earlier. In a certain way, it can be seen as a continuation of the development and research activities in aviation medicine. And here, Europe and especially Germany had a long-lasting leading position, above all via the activities in Berlin. Already in 1912, Nathan Zuntz had published his book "Zur Physiologie und Hygiene der Luftfahrt" (Zuntz 1912; On the Physiology and Hygiene of Aviation; see also Gunga and Kirsch 1995), and Zuntz can be seen rightfully as the father of aviation medicine. Already at the end of the 1930s, scientists in Berlin around Otto Gauer and Hubertus Strughold used centrifuges in their attempts to mimic the conditions existing in airplanes and tested the effects of high g-loads on human physiology.

In addition to the already mentioned well-known transfer of rocket specialists after World War II to the US, the Americans also took the chance to gather German scientists and medical doctors familiar with aviation medicine in Heidelberg. They were tasked to discuss their expertise and to write down their findings. The results were published in 1950 in two volumes with the title "German Aviation Medicine in World War II" (USAF 1950). Interestingly, there was already a contribution by O. H. Gauer and H. Haber with the title "Man under gravity-free conditions" (Gauer and Haber 1950), in which the authors extrapolated the results of their centrifuge experiments in the direction of microgravity. Problems of the cardiovascular, the vestibular system, the inner ear as well as of the muscle and bone system were identified as major challenges for human spaceflight. Indeed, many of the assumptions of Gauer and Haber could be confirmed in later space experiments, although we know today, of course, that a simple extrapolation form hyper-g via 1 g gravity on Earth to microgravity does not work in many cases.

Also, shortly after World War II, prominent German aviation physiologists such as Hubertus Strughold, Konrad Buettner, and the brothers Fritz and Heinz Haber were brought to the Randolph Airforce School of Aviation Medicine near San Antonio, Texas in order to work on the physiological consequences of spaceflight on the human body. In fact, it was Strughold who in 1948 introduced the term "space medicine," and only 1 year later, in February 1949, the first department of space medicine in the world was established at the Randolph Airforce School under Strughold's leadership (Strughold 1952). So, not only German space engineers such as Wernher von Braun but also German life scientists had a substantial say in the development of the US space activities (for more details on history see Harsch 2008; Reinke 2007; Ruyters 2008).

Nevertheless, more than two decades had to pass before European scientists could perform their own space experiments as cooperative projects with the US or with the Soviets. From the 1980s onwards, experiment possibilities could be slowly presented on a more regular basis. This became possible in the context of the American invitation to Europe to cooperate in the frame of the space shuttle development, precisely with the European decision to develop and contribute the Spacelab (see below and Chap. 2).

1.5   Organizational and Programmatic Aspects in Europe

After World War II, Americans and Russians had an enormous head start on Europe and Germany not only in terms of space technology but also in space research. In post-war Europe, the organizational and programmatic foundations of space-based research simply did not exist. This certainly does not come as a surprise since Europe was much more involved in trying to solve economic problems and challenges these days. Indeed, the West European nations soon began the process of combining their economic forces. The European Economic Community established in 1957 with Belgium, the Netherlands, Luxembourg, France, Italy, and Germany as founding members represented the second-largest economic power in the world. Despite that fact, Europe did not play any role in the demonstration of technical leadership in space at that time. This was going to be changed. Consequently, a desire was created in Europe to participate in the next steps in the manned exploration of space.

1.5.1   Foundation of ESA and Its Manned Spaceflight and Microgravity Programs

In fact, two European space organizations were established in 1962, the Space Research Organization (ESRO) and the European Launcher Development Organization (ELDO) to define and built-up European capabilities and activities in space. With a progressive expansion of their activities and capabilities, Europe was in a position by 1973 to accept an invitation from NASA to participate in an international Post Apollo Program aiming at the development of the Space Shuttle System. This important decision marked the beginning of the manned spaceflight program within

Fig. 1.1 Crew of STS-9 including German astronaut Ulf Merbold (second from right) which took off on the first Spacelab mission aboard Space Shuttle Columbia in 1983. (© ESA/NASA)

Europe, and it also led to the merger of ESRO and ELDO into a full European Space Agency in 1975. The decision included also the development of a European launcher system later called Ariane and a series of telecommunication satellites. These elements together with the Space Science Program of the former ESRO were to form the core activities of ESA. Founding member states of ESA at that time were Belgium, Denmark, France, Germany, Italy, the Netherlands, Spain, Sweden, Switzerland, and the United Kingdom (for more details see Seibert 2001).

The joint European-American selection of Spacelab as the European element of the STS (Space Transportation System) was mainly due to the US national security concerns about overlapping technology developments, and at the same time to the European desire to develop an easily identifiable element of the STS. The first Shuttle/Spacelab mission took place on November 28, 1983, with Ulf Merbold as the first ESA and non-American Shuttle astronaut (Fig. 1.1).

The selection of the German Ulf Merbold was a strong indication of the role Germany played in manned spaceflight programs in Europe from the beginning. Germany was the biggest shareholder in the respective ESA programs, German industry had the leadership in the development and construction of Spacelab. The contribution of Spacelab to the US Shuttle system turned out to be the point of entry into the most complex aspects of space flight—astronautics. At the end, it was owing to the pertinacity of Germany that Europe would succeed—through Spacelab—in gaining access to manned missions for its astronauts via bi- or multilateral cooperation without having a launcher of its own. Today, about 40 years later, this is still the valid and successful strategy for Europe also in the era of the International Space Station.

Moreover, it must be stressed at this point that the contribution of Spacelab was not limited to the infrastructure as such. The strong engagement of ESA and especially of Germany in manned spaceflight also led to the formation of a new scientific community busy with new scientific research activities in life and physical sciences in Europe. To enable this research in space under microgravity conditions so-called multiuser experiment facilities had to be developed. Therefore, specific microgravity programs were established at ESA from 1982 onwards (for details on the programs of the 1980s and 1990s, see Seibert 2001). ELIPS (European Life and Physical Sciences Program) started in 2001 with 15 participating member states making use of a number of ground-based microgravity platforms such as the Bremen Drop Tower, parabolic airplane flights and sounding rockets, as well as of satellites, the Space Shuttle System, and especially the ISS. At the ministerial conference in 2016, ELIPS was replaced by SciSpacE (Science in Space Environment), now embedded in the European Exploration Envelope Program (E3P)—that still exists today. The SciSpacE program involves over 1200 scientists from 19 participating ESA member states and over 60 industrial R&D partners caring especially for advancing Europe's knowledge base by the successful utilization of the various research platforms—first of all the ISS—and for the preparation of the next steps of exploration (for details see https://www.esa.int/Science_Exploration/Human_and_Robotic_Exploration/Research/European_user_guide_to_low_gravity_platforms).

1.5.2   European National Programs in Space Life Sciences

In parallel to the development of ESA—and possibly at least to some extent due to the difficulties and delays associated with the creation of the ESA microgravity program—several of the ESA member states had been developing their own national programs in space life and physical sciences since the 1970s. This was especially the case for Germany, and to a lesser extent for France and later for Italy (see Seibert 2001 for further details).

1.5.2.1   A Short History of the German Space Life Sciences Program: DLR in Its Role as Space Agency

From today's viewpoint, it seems extremely surprising that as early as 1960, i.e., already 2 years before the establishment of ELDO and ESRO as European space organizations, science and politics in Germany began to address the issue of space research: the German Research Foundation first analyzed the situation in its 1961 "Memorandum on the status of space research" (Reinke 2007; remark: Germany stands for West Germany, i.e., the Federal Republic of Germany, in this historic chapter). In January 1961, Chancellor Konrad Adenauer decided to put all space-related activities into the hands of what was then called the Atom-Ministerium or Nuclear Research Ministry and later transformed into the Ministry of Scientific Research in 1963. Within that ministry, a "department of space studies, space flight research and technology" was established, wherein a special "unit on research grants for space studies, the physics of far-Earth and near-Earth space, space biology, and satellite research" was put in charge of exploring all the options from a programmatic and scientific point of view. Germany's first space program was presented in 1967 and countersigned by the Federal Government. Like the second space program of 1969 however it mentioned life sciences research only as an afterthought.

This changed—at least to some extent—with the Spacelab decision. In the frame of preparing the development and utilization of Spacelab the German government was not only involved in supervising the activities within the respective ESA programs but also initiated comprehensive national activities. In fact, German authorities created preparatory and complementary activities such as the successful sounding rocket program TEXUS with its first flight in 1977 (see Chap. 2 for more details). This seemed the more necessary, since Germany took over the responsibility to conduct the two national Spacelab missions D-1 in 1985 and D-2 in 1993, in cooperation with the US, when—due to the escalating flight costs—the ESA Member States were unwilling to fund further Spacelab missions after the initial joint NASA/ESA Spacelab-1 mission. Since Germany had already contributed 53% of the costs of the Spacelab development program of ESA, a continuation of Spacelab flights as part of the German national program seemed to be the most logical next step. International partners were invited to join, and at the end ESA participated in D-1 (1985) with 38% of the payload and the Dutch astronaut Wubbo Ockels with Reinhard Furrer and Ernst Messerschmid being the German astronauts. Hans Schlegel and Ulrich Walter were chosen as astronauts for the D-2 mission (1993), in which ESA had a share of 25% participation of the payload.

For these German missions, further experiment facilities had to be developed by industry based on the experience from the first Spacelab mission. In addition, funding of scientists participating in this research had to be guaranteed according to rules and procedures that had to be established. These missions formed the basis for Germany's leading role in manned spaceflight and microgravity research in Europe to this day, and ensured that the German Space Agency, German space industry, and German scientists are judged as reliable and important cooperation partners for their counterparts around the world.

Despite TEXUS and the successful first Spacelab mission however the true importance of microgravity research in life sciences was not appreciated in politics and government in Germany. Even in the fourth German space program of 1983, it still appeared rather inconspicuously as a sub-program of extraterrestrial basic research entitled "biological-medical research in extraterrestrial space." It was only in the late 1980s—maybe in part due to the results of the successful FSLP (First Spacelab) mission of 1983 that were published in a special issue of the high ranking scientific journal "Science" in July 1984 (https://science.sciencemag.org/content/sci/225/4658/127.full.pdf)—that the foundations were laid for an autonomous life sciences program: Independent expert bodies developed program papers for three research areas, biology, biotechnology, and medicine, which were published for internal planning purposes in 1989. These program papers defined a number of overarching issues and research questions and discussed ways and means of addressing them. The overall goal defined in 1989 is still valid today: "To acquire scientific knowledge and develop new application potentials through biomedical research under space conditions, particularly in weightlessness."

This goal is also supported in the latest space strategy document of the German government "Making Germany's space sector fit for the future: The space strategy of the German Federal Government" published in 2010 (https://www.dlr.de/rd/en/

Portaldata/28/Resources/dokumente/Raumfahrtstrategie_en.pdf). On this basis, three focal objectives have been established with the support of external experts within the German life sciences program (see also Ruyters and Friedrich 2006):

- Exploring Nature with research and development activities in gravitational, radiation, and astrobiology as well as in biotechnology.
- Improving health with research and development activities in the various areas of human physiology including countermeasure development.
- Enabling exploration with research and development activities especially in the areas of human health and performance and bioregenerative life support systems.

Some of the scientific questions raised in the early days have been answered by now, while others have been added to the list. In the years and decades to come, exploratory long-range missions to the Moon, Mars, and other remote destinations will confront life sciences research in space with new challenges—but more about that in the outlook Chap. 5.

From an implementation point of view, the German government had delegated all activities already of the Spacelab era to an institution that is nowadays known as the German Aerospace Center (DLR; in German: Deutsches Zentrum für Luft- und Raumfahrt e.V.). Historically, today's DLR is the result of a series of organizational changes across the twentieth century (for more details see Reinke 2007). Its oldest predecessor organization was established by Ludwig Prandtl in Göttingen in 1907. After World War II, the "Gesellschaft für Weltraumforschung" (GfW; "Society for Space Research") that had already existed between 1935 and 1945 was reestablished. It was—by the way—the institution that in 1949 put a resolution to foreign astronautical associations that led to the founding of the International Astronautical Federation (IAF). It also helped—at the beginning with Hermann Oberth as honorary chairman—to organize the International Astronautical Congress from 1950 onwards, which still plays an important international role in astronautics with its 71st congress held due to the COVID-19 pandemic in October 2020 as a CyberSpace Edition.

In 1969, the Deutsche Forschungs- und Versuchsanstalt für Luft- und Raumfahrt (DFVLR; "German Test and Research Institute for Aviation and Space Flight") was formed through the merger of several institutions. In 1989, DFVLR was renamed Deutsche Forschungsanstalt für Luft- und Raumfahrt (DLR; "German Research Institute for Aviation and Space Flight"). In the same year, the Deutsche Agentur für Raumfahrtangelegenheiten (DARA; "German Agency for Space Flight Affairs") was created with the idea to form an independent German Space Agency. Already in 1997 however DARA was "reintegrated" into DLR to build the new DLR, the German Aerospace Center, as we know it today.

While the headquarters are in Cologne, its more than 50 institutes and facilities are spread across 34 locations throughout Germany, with offices in Brussels, Paris and Washington, D.C., and Tokyo coming on top. DLR currently employs over 8200 people. In national and international partnerships, DLR is engaged in a wide portfolio of research and development projects ranging from aeronautics, space-flight, and transport to energy research. The Institute of Aerospace Medicine in

Cologne is the DLR research entity involved in life sciences research and also in operational spaceflight activities. Its first predecessor had been established in Berlin already in 1934, moved to Bonn in 1947, and then in 1969 to the newly built premises in Cologne-Porz.

In addition to conducting its own research projects, DLR also acts as the German space agency located in Bonn. On behalf of the German government, DLR is responsible for defining, deciding, and supervising—together with the other member states—the various ESA programs as well as establishing and implementing the national program. Among the various elements of the national program, the so-called Microgravity Research and Life Sciences Program combines the activities in physical and materials sciences as well as in space life sciences covering gravitational, radiation, and astrobiology, biotechnology, and human physiology, health, and performance. All in all, it is fair to say that this strategy of trying to play a leading role in the respective ESA programs and at the same time establishing a strong national program focusing on bilateral cooperation with basically all space faring nations turned out to be the basis for the success stories described in this volume.

1.5.2.2 France, Italy, and Sweden: Partners with Different Competencies

The financial envelope for France's space activities consisting of its contribution to ESA and the national space projects has made it the largest investor in space in Europe over many years—only recently Germany has taken the lead also in this respect. All French space programs are implemented by the French space agency CNES.

France's leading position in terms of European space expenditures is not however fully reflected in its funding for life and physical sciences research: Its contribution to the various Manned Spaceflight and Microgravity Programs of ESA has—on average—only been about half that of Germany. The focus of the French space activities is more on the ARIANE and the ATV programs. Nevertheless, a substantial program in space life sciences could develop especially in the 1980s and 1990s. In this period seven cooperative manned missions with the Soviet Union and Russia, respectively, to the space stations Salyut-7 and MIR have been conducted between 1982 and 1999 and led to an important role of France in space medical research. The investments of France for the International Space Station were, on the contrary, rather meager, which also led to a decreasing importance of the accompanying activities in the national program. It remains to be seen how France will position itself in the upcoming era of exploration.

Italy with its space agency ASI has become more active over recent years especially due to its national contribution to the ISS infrastructure in the form of the MPLM (Multi-Purpose Logistics Module). With this contribution, it has acquired additional utilization rights of the ISS over and above the utilization via ESA.

In Sweden, the national program in microgravity research has evolved around its sounding rocket activities. Through the Swedish Space Corporation, the Swedish national program provides support for stratospheric balloon and sounding rocket missions including REXUS, MASER, TEXUS, MAXUS (discontinued after MAXUS-9 in 2017), and MAPHEUS, which are launched under contracts of ESA and DLR from the Esrange Space Center. Europe's largest civilian space center is located near Kiruna in the very north of Sweden above the Arctic Circle; it also accommodates one of the world's largest civilian satellite ground stations (https:// www.sscspace.com/ssc-worldwide/esrange-space-center; for more information on the European national programs, see Seibert 2001).

1.5.3 Space Research in the German Democratic Republic Between 1957 and 1990: Participation in the Interkosmos Program

A summary on the historical development of space research, especially in life sciences in Europe, would not be complete without including the activities of the former German Democratic Republic (GDR). Also, in "the other Germany" the years after World War II were characterized by deconstruction and a loss of all aerospace activities with East Germany taking even longer to recover than West Germany (for more details see Reinke 2007).

The official historical record of the GDR's involvement in astronautics started with the flight of Sputnik 1 in 1957, which it tracked—like many other countries—from its own observatories using optical and radio techniques. The cooperation between the GDR and the Soviet Union first took place between the respective academies of sciences and was initially restricted to the observation, documentation, and evaluation of Soviet missions.

In June 1960, the Astronautical Society of the GDR was established; in August of the same year, it was accepted as member of the International Astronautical Federation (IAF). This was more than a surprise—in a certain way, it was a political sensation for two reasons: As a first upset, it was achieved against the will of the state apparatus of GDR, which still equated space flight with building missiles and not with promoting research and utilization of space for peaceful purposes—the stated goal of the IAF. Secondly, it was—in contradiction with the so-called Hallstein doctrine—a successful attempt to win international recognition in the process of establishing two German states and as such directed against the Western countries. However, at the end, the West German IAF delegate was the only one to vote against the acceptance of GDR as a full IAF member with voting rights.

Like the space activities of the Federal Republic of Germany life sciences or microgravity research did not play any role at the beginning in the collaboration between the GDR and the Soviet Union. This changed with the establishment of the INTERKOSMOS program in 1967, in which the conditions of cooperation between

Fig. 1.2 Happy faces after landing. Cosmonauts Walery Bykowski (left) and Siegmund Jähn (right) from the German Democratic Republic returned to Earth from the third Interkosmos Mission. (© Bundesarchiv Bild 183-T0905-107/CC-BY-SA 3.0)

the Soviet Union and the member states Bulgaria, GDR, Cuba, Peoples Republic of Mongolia, Poland, Romania, Czechoslovakia, and Hungary were regulated: The Soviet Union would provide—free of charge—the carriers, rockets, satellites, and stations on Earth in exchange for access to all data collected from the experiments by the partner countries. Respective working groups for structuring the activities were set up for space physics, meteorology, communication, biology and medicine, and Earth remote sensing.

In addition to conducting experiments on cells, plants, and animals in a number of reentry satellite flights, the outstanding activity became the cooperation in biology and medicine onboard the Soviet space stations. Between 1978 and 1988 cosmonauts from all ten INTERKOSMOS partner countries took part in human spaceflight missions—among them the first German to find his way into space, Sigmund Jähn. Following a 2-year training program in Star City near Moscow, which was kept totally secret in his own country, Jähn launched on August 26, 1978, onboard Soyuz 31 from Baikonur to the Salyut 6 space station for a 7-day mission (Fig. 1.2). Technical, medical, and biological experiments were in the focus, among them studies on human time perception as well as on the sensitivity of hearing and taste in microgravity.

Space research activities in East Germany did not stop because of the unification in Germany in 1990. Although the integration into the West German research and industrial communities turned out to be more difficult than expected, the close ties of the "other Germany" to Moscow—which for many years had been viewed with skepticism in the West—proved to be a benefit in the field of astronautics. Germany thus became a preferred cooperation partner with Russian authorities for the use of the space station MIR and later for utilizing the Russian part of the International Space Station ISS. This is not the least due to the high appreciation of Sigmund Jähn in Russia as well as in Western Europe. Already in the preparation of the Russian-German mission MIR'92, Jähn was hired as an advisor for the German space agency DLR. Interestingly, the connection was established—at least to some extent—by the first West German astronaut, Ulf Merbold, who had met Jähn already in 1984 at a conference in Salzburg. Both had been born in a small area in the Southeast of the former GDR some 20 km apart from each other with Merbold moving to West Germany in 1960 and Jähn remaining in the GDR. Later Jähn also supported ESA in the preparation of its EUROMIR missions in 1994 and 1995. In November 2019, Jähn died at the age of 82—highly appreciated by space authorities, colleagues, and astronauts and cosmonauts in East and West.

References

Gauer O, Haber H (1950) Man under gravity-free condition. In: The Surgeon General (ed) German aviation medicine World War II, vol 1. USAF, US Government Print Office, Washington, pp 641–644

Gunga HC, Kirsch KA (1995) Nathan Zuntz (1847–1920) – a German pioneer in high altitude and aviation medicine. Aviation Space Environ Med 66:168–176

Harsch V (2008) Anfänge der Flugmedizin in Deutschland nach 1945. In: Wirth F, Harsch V (eds) 60 Jahre Luft- und Raumfahrtmedizin in Deutschland nach 1945. Rethra, Neubrandenburg, pp 1–8

Oberth H (1923) Die Rakete zu den Planetenräumen (English translation 2014: The rocket into interplanetary space). De Gruyter, Oldenbourg

Reinke N (2007) The history of German space policy - ideas, influences, and interdependence 1923–2002. Editions Beauchesne, Paris

Ruyters G (2008) Raumfahrtmedizin in der Bundesrepublik Deutschland. In: Wirth F, Harsch V (eds) 60 Jahre Luft- und Raumfahrtmedizin in Deutschland nach 1945. Rethra, Neubrandenburg, pp 91–122

Ruyters G, Friedrich U (2006) Gravitational biology within the German space program: goals, achievements, and perspectives. Protoplasma 229:95–100

Seibert G (2001) A world without gravity – research in space for health and industrial processes. ESA SP-1251

Strughold H (1952) From aviation medicine to space medicine. J Aviat Med 23:315–318

USAF (1950) German aviation medicine World War II, 2 Vols. USAF, US Government Print Office, Washington

Zuntz N (1912) Zur Physiologie und Hygiene der Luftfahrt. Springer, Berlin

Chapter 2
A Long Way for Europe and Germany: From Apollo 16 to the International Space Station ISS

Abstract When Apollo 16 took off for the Moon in 1972, the Federal Republic of Germany launched its first life science research project in space: the BIOSTACK experiment to study the intensity and composition of cosmic radiation became part of the Apollo missions. The TEXUS rocket programme was initiated in 1976 to pave the ground for the European space laboratory Spacelab, whose 1983 maiden flight on an American Space Shuttle raised microgravity research to an entirely new quality. The German Spacelab missions D-1 and D-2 followed and further cooperation between Europe and the US was realized during the Shuttle/Spacelab era in the frame of the so-called IML model. The German-Russian MIR'92 and MIR'97 missions, two EUROMIR missions by ESA, and several CNES missions to MIR provided Europe's scientists with further attractive research opportunities, as did parabolic flights on an Airbus A300 from the late 1990s onward. A new chapter has been opened by the ISS and its European Columbus laboratory. In addition, terrestrial simulation capabilities for biology, as well as bed rest, isolation, and confinement studies for research in human physiology and psychology, prepare and complement life sciences research in space.

Keywords Apollo · Bremen Drop Tower · Isolation and bed rest studies · ISS · MIR space station · Parabolic Flights · Reentry satellites · Simulation facilities · Space Shuttle/Spacelab · TEXUS sounding rockets

2.1 Flight Opportunities for Space Life Science Research in Europe

Sixty years after the first manned missions of the Soviet Union and of the United States, there is still no autonomous European access to space for humans. In contrast, already from the 1960s, the USA had developed its Mercury, Gemini, and Apollo programmes as well as its first space station, Skylab, and later the Space Transportation System STS (Launius 2020). In parallel, the Soviet Union had developed its Vostok, its space stations Salyut and later MIR, and Russia is still launching astronauts and cosmonauts to the ISS by its Soyuz spacecrafts (pbs.org/spacestation/ station/Russian.htm). Thus, the situation in Europe remains different from that of the US and Russia, and meanwhile also of China. In order not only to fly European astronauts, but also to enable research in biology, human physiology, and medicine in space for European scientists, extensive international cooperation was mandatory over decades—and still is (Table 2.1).

Europe, and here again especially Germany, probably took the best consequence out of this situation and developed—in parallel to the Spacelab decision—what is sometimes called with some understatement the "small flight opportunities." These are the drop tower in Bremen, the sounding rockets program, and parabolic airplane flights. The utilization of reentry satellites via cooperation with the Soviet Union/Russia and also China provides additional opportunities.

All in all, it is fair to say that these developments plus the successful cooperation in human spaceflight with the US and the Soviet Union or Russia, respectively, put ESA as well as DLR into the position to provide the broadest portfolio of flight opportunities worldwide for life and physical sciences research in microgravity. A short summary of these various flight opportunities may therefore be justified (see also Ruyters and Friedrich 2006; Ruyters and Preu 2010; Ruyters 2020; further information is available on the respective websites of ESA and DLR).

2.1.1   The Bremen Drop Tower

Unique in Europe, the Bremen Drop Tower in northern Germany is a scientific facility for short-term microgravity experiments (Fig. 2.1). Since 1990, literally daily scientific and technological experiments are being performed making use of the 4.7 s of weightlessness that is provided by each drop. To achieve this, the experimental hardware developed and built by the scientists with the support of the experienced engineers of the drop tower team is installed into the drop capsule. The drop tube of 120 m inside the tower of a height of 146 m is evacuated before the drop by high-powered vacuum pumps. Immediately after the release, the capsule is in a free-fall situation and the gravity level is reduced to almost zero. Braking of the falling capsule is then being achieved within the deceleration container filled with polystyrene pellets of 5 mm diameter. Since 2004, doubling of the μg-time to 9.3 s is achieved by using the catapult system (www.zarm.uni-bremen.de/en/drop-tower).

While in the early years the vast majority of experiments performed was in physical sciences, especially in fluid physics, combustion, rheology, and materials research, recently also biologists increasingly make use of the facility. This is on one hand due to the fact that more sophisticated analysis techniques became available. On the other hand, a shift of the research focus towards the very early and fast responses of cells, tissues, and organisms to changed gravity conditions during signal transduction made drop tower experiments highly attractive. In fact, the gravity-induced absorption changes in the fungus Phycomyces and in higher plants

Table 2.1 Flight Platforms accessible for European Space Research

	Drop Tower	Parabolic Flight	Sounding Rockets	Space Shuttle	Satellite	Space Station
Microgravity duration	4.7/9.3 s	31 parabola @ 22 s, 3 FDs	4–12 min	14 days	Up to 20 days	Permanent
Microgravity quality	$<10^{-5}$ g	$\sim10^{-2}$ g	$<10^{-4}$ g	$\sim10^{-3}$ g	$\sim10^{-6}$ g	$\sim10^{-4}$ g
Experiment manufacturer[a]	E	E	I	I	E/I	I
Experiment operation[a]	Auto	E	Auto/E via telecommand	Auto/A/E via telecommand	Auto/E via telecommand	Auto/A/E via telemetry/ telecommand
Latest access	30 min	10 min	60 min	17 h	5 days	17 h
Early retrieval	30 min	1 min	30 min	3 h	2 days	3 h
Lab at launch site	Yes	Yes	Yes	Yes	Yes/no	Yes

[a]E, experimenter; I, industry; auto, automatically; A, astronaut; figures represent typical/average values; FD, flight day

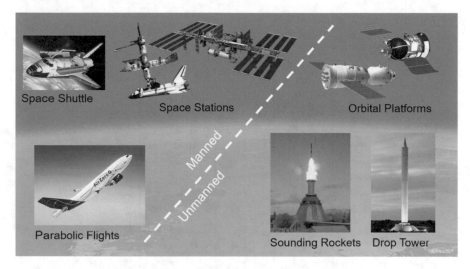

Fig. 2.1 Flight platforms for European space life science research. (© M. Braun)

such as Zea recorded in drop tower experiments after 20 ms of microgravity belong to the fastest responses ever measured (Schmidt 2010; Schmidt and Galland 2004).

2.1.2 *Research on Parabolic Flights Using Aircrafts: 22 s of Free Fall*

Another opportunity to achieve free-fall conditions is by flying an aircraft in a special maneuver (Fig. 2.1). For many years, in Western Europe a Caravelle aircraft (1989–1995) and an Airbus A300 ZERO-G (1997–2014) were used for these flights. Since 2015, a former Chancellor aircraft of the Federal Republic of Germany, an Airbus A310, has taken over the task to perform parabolic flight campaigns for ESA, CNES, and DLR (for details see Pletser et al. 2015). These parabolic flights with airplanes are the only opportunity where the scientists can participate in the flights and operate their experiments and at the same time experience weightlessness by themselves. Except for the International Space Station ISS, it is currently the only possibility to regularly perform medical and physiological studies on humans under microgravity conditions.

A campaign normally comprises three flight days, with 31 parabolas being flown each day. The aircraft steeply ascends from the horizontal at an angle of up to 50° and then throttles back its turbines to enter a parabolic path that corresponds to a ballistic trajectory. At that time, the plane and everything and everyone onboard is in free fall for about 22 s. Before and after the microgravity phase however passengers, biological samples, and equipment are exposed to nearly double the Earth's gravity for about 20 s. Being certainly a challenge for scientists and infrastructure, also these rapid changes in gravity conditions yield important insights into the responses of biological specimen as well as into gravity-affected physiological processes in humans.

Space agencies have been relying on parabolic flights for decades. Until the 1980s however these flights mainly served to train astronauts and test technologies and equipment in preparation for later human missions. This changed in the 1990s, when their potential as an independent and stand-alone flight opportunity for experiments in alternating hyper- and microgravity conditions was increasingly recognized, helped along by the development of new analytical technologies and the investigation of rapid effects in the perception and transduction of gravity. While ESA is regularly organizing parabolic flights since the 1980s, cooperating with NASA at first and then, from 1989 onwards, with Novespace, a subsidiary of the French space agency CNES, the DLR Space Administration provides this research opportunity to German scientists since 1999.

Between 1999 and 2020, DLR ran 35 parabolic flight campaigns under the national space program; 74 campaigns were organized by ESA and 59 by CNES. About half of the projects in life sciences focussed on the signaling pathway of the perception and transduction of gravity in microorganisms, protozoa, fungi, fish, algae, and plants at the cellular and molecular level. The other half was concerned with rapid changes in various physiological systems of the human body, such as the movement of body fluids, cardiovascular regulation, neuro-physiological and neurocognitive changes, and movement coordination. Since a few years, also parabolic flight campaigns providing Moon (0.16 g) or Mars (0.3 g) gravity conditions are offered to the scientists as joined endeavors of ESA, CNES, and DLR (Pletser et al. 2012). In 2019, also NASA scientists participated in one of these parabolic flight campaigns—more is probably to come.

2.1.3   TEXUS, MASER, MAXUS: The European Research Rocket Family

Established in 1976 by the German Ministry of Research and Technology immediately after the decision to develop and build Spacelab as European contribution to the US Space Transportation System (STS), the TEXUS program (Technologische Experimente unter Schwerelosigkeit—Technological Experiments under Microgravity; Fig. 2.1) was launched to prepare a meaningful scientific utilization of Spacelab. For such a preparation program, a DLR study had clearly favored ballistic rocket flights (Franke 2007). Already on December 13, 1977, the first TEXUS rocket providing 6 min of microgravity took off from the polar circle to carry research apparatus to space and back. The rocket's launch from Kiruna in the north of Sweden was the curtain-raiser of a scientific program that extended successfully over decades. Today, the flight opportunities offered by TEXUS support stand-alone research projects as well as help pave the way for experiments on the International

Space Station (ISS). Since the maiden flight in 1977, 76 life sciences experiments have successfully been conducted on 56 TEXUS campaigns covering many fields of biology like biotechnology, pharmacology, and gravitational biology comprising many aspects of plant and animal physiology, molecular biology, developmental biology, microbiology, and immunology. Interestingly, also the first experiment on protein crystallization in microgravity worldwide was executed on TEXUS—on TEXUS-3 in 1981 namely—yielding large crystals of the enzyme beta-glucosidase thus paving the way for establishing a completely new research field at that time (see Sect. 3.3.4).

In 1982, ESA joined the DLR Space Administration in providing such flights for selected experiments and later tasked industry to develop a system that would provide even 12–13 min of microgravity, the MAXUS project. Its first successful flight was launched in November 1992. Since 1987, the company Airbus Defense & Space has been responsible for coordinating the program. DLR's mobile rocket base (MORABA) and the Munich-based company Kayser-Threde (today OHB Bremen) are involved as well. Together with the Swedish MASER, TEXUS and MAXUS have formed the European sounding rocket family—more recently complemented by the newly established MAPHEUS program for material physics experiments and the German-Swedish REXUS program for student experiments. A few years ago, the MAXUS project was put on hold by ESA due to its high costs.

On their ballistic flights, the two-stage 13-m solid-fuel TEXUS rockets reach a peak altitude of about 250 km, providing approximate weightlessness for about 6 min before returning to Earth on a parachute. The tip of the rocket houses up to five experiment modules as well as recovery and data transmission systems. During the microgravity phase, researchers can monitor their experiments in real-time and may even directly change parameter settings by telecommand. Scientific data are either transmitted by telemetry during the flight or stored onboard and secured after payload recovery. Due to the very late access to the biological samples and the early recovery of the samples within 1 h after flight, sounding rockets are especially attractive for life sciences experiments. Studies on signal transduction mechanisms in gravitaxis of mobile cells, in gravitropism of plants as well as in immune and cancer cells have come into focus with the recent availability of the innovative FLUMIAS experiment facility for life cell imaging (see Sect. 4.5).

2.1.4   Use of Reentry Satellites in Cooperation with Russia and China

For many years, so-called reentry satellites—particularly FOTON and BION—have been used by ESA, DLR, and other ESA member states for life sciences experiments in cooperation with Russia (Fig. 2.1). The capsules provide high-quality microgravity, the flights having usually a duration of a few days until several weeks. While West-European scientists were offered the opportunity of using these Russian satellites since the late 1980s—via ESA since 1987, in the DLR program since 1989—scientists from Eastern Germany already conducted many experiments between 1979 and 1989 in the frame of the INTERKOSMOS program.

The technique behind the Russian satellite missions is rather old, but reliable. In fact, the FOTON capsule was developed on the basis of the Wostok spacecraft that took Gagarin into space in 1961. In the meantime, several new developments and capabilities like solar panels have been implemented. The capsules are launched with a Soyuz rocket from Baikonur cosmodrome in Kazakhstan, the landing then is on Russian territory close to Kazakhstan. The experiments are performed automatically, a limited telemetric access allows monitoring and tele-commanding of the experiments. About ten flights with ESA or DLR participation have been performed, covering a wide range of scientific disciplines in physical as well as in life sciences. In life sciences, the focus has been—in addition to radiation, astrobiology, and protein crystallization—in gravitational biology, especially on gravitaxis, gravitropism, gene expression, and on developmental aspects of small animals like flies. Recently, also tests of aquatic bioregenerative life support systems including research on fluxes of energy and matter between the different biological components have been conducted by German scientists.

Interestingly, in the timeframe between 1989 and 1991, some additional satellite projects were performed in an early cooperation between DLR and China. Four so-called COSIMA missions (Crystallization of Organic Substances in Microgravity for Applied Research) took place providing microgravity for research on protein crystallization. In these times, the Chinese reentry capsules with the German experiments onboard were launched on top of a Long March rocket from Jiuquan space center in Inner Mongolia. It should take another 20 years until the next joint German–Chinese space project in life sciences, SIMBOX, was launched on the Chinese reentry capsule Shenzhou-8 in 2011 (see Sect. 2.1.8).

2.1.5   Spacelab: A Milestone in European and German Microgravity Research

As already mentioned in Chap. 1, Spacelab was Europe's contribution to the American Space Shuttle in the frame of the space transportation system (STS) program (Fig. 2.2). In fact, Spacelab was not just a single piece of infrastructure, but a suite of elements, consisting of a pressurized module in the cargo bay of the shuttle that served as a laboratory environment, unpressurized platforms—so-called pallets—and various other H/W components that were reconfigured to meet the specific requirements for each Space Shuttle mission, thereby turning the shuttle into an efficient laboratory platform for scientific research in space. The very first Spacelab project (FSLP), Spacelab's first journey into space, was performed in 1983. The first German Spacelab mission, D-1, followed in 1985 with Reinhard Furrer and Ernst Messerschmid as well as ESA astronaut Wubbo Ockels onboard.

Fig. 2.2 ESA's Spacelab module in the payload bay of Space Shuttle Columbia was the workplace of the German ESA astronaut Ernst Messerschmid and his STS-61-A fellow crew members during the D1 mission in 1985. (© NASA)

Table 2.2 European cosmonauts on US Shuttle/Spacelab Life & Physical Sciences Missions

Name	Participating organization	Mission	Date
U. Merbold (D)	ESA	Spacelab-1	28.11.–8.12.1983
R. Furrer (D)	DLR	Spacelab D-1	30.10.–6.11.1985
E. Messerschmid (D)	DLR	Spacelab D-1	30.10.–6.11.1985
W. Ockels (NL)	ESA	Spacelab D-1	30.10.–6.11.1985
U. Merbold (D)	ESA	IML-1	22.–30.1.1992
H. Schlegel (D)	DLR	Spacelab D-2	26.4.–6.5.1993
U. Walter (D)	DLR	Spacelab D-2	26.4.–6.5.1993
J.-F. Clervoy (F)	ESA	6th Shuttle/MIR	15.–24.5.1997
J.-L. Chrétien (F)	CNES	7th Shuttle/MIR	25.9.–6.10.1997
P. Duque (E)	ESA	Spacehab	29.10.–7.11.1998

Next, after some delays due to the Challenger accident, D-2 was launched in 1993 with the two German astronauts Ulrich Walter and Hans Schlegel (Table 2.2). While the public showed particular interest in these missions because of the German astronauts, FSLP, D-1, and D-2 also received great attention for the multitude of scientific results in materials science, biology, and medicine thus representing highlights in the German microgravity research program. Based upon this close German American cooperation, excellent relations with NASA evolved which enabled German life sciences experiments to be run on other shuttle flights as well. This was achieved for DLR as well as for ESA—and to a lesser extent also for other partners—within a framework of cooperation later known as IML model (International Microgravity Laboratory). Under this agreement the European partners developed and provided so-called multiuser facilities necessary for the scientific experiments, NASA took over the cost of the flights and the orbital operations of these facilities on the Shuttle/Spacelab missions. The utilization of the facilities was then to be shared (50/50) between American and European scientists. With this favorable arrangement, several NASA Shuttle/Spacelab missions were opened for European researchers and astronauts after IML-1 in 1992, among them IML-2 in

1994, USML-2 (US Microgravity Lab) in 1995, LMS (Life and Microgravity Sciences) in 1996, and finally Neurolab in 1998, the last Spacelab mission.

Neurolab—totally dedicated to neuroscience research—was special also in another context. It saw contributions not only from NASA and ESA but also from DLR and other space agencies such CSA, CNES, and JAXA. In 1991, these agencies had established a framework to foster cooperation and coordination in space life sciences by creating the International Space Life Sciences Working Group (ISLSWG); later also the Italian and the Ukrainian Space Agencies joined the group (www.nasa.gov/directorates/heo/slpsra/islswg.html). The whole Neurolab mission was planned and organized by ISLSWG: the distribution of experiment facilities was discussed and agreed upon among the partners, and finally the scientific experiments were jointly solicited, peer-reviewed, and selected. Neurolab thus turned out to be a great model for the way science and research should be successfully organized for ISS. Germany via DLR had a substantial share in this outstanding endeavor and provided the aquatic bioregenerative life support system CEBAS, the BOTEX incubator for studies on the development of the gravity sensor system in crickets as well as the LBNP (Lower Body Negative Pressure System) supporting cardiovascular research of scientists of the so-called autonomous function team. In addition, visuomotor coordination by astronauts in space was analyzed in a joint Canadian/German project (Buckey and Homick 2003).

In total, more than 140 scientific projects under German leadership, implemented mainly by DLR in its national program but also via the German membership in ESA, have been performed over the years in the Space Shuttle era between 1981 and 1998, 125 of them within the Spacelab module. Different from the utilization of the MIR space station, but also different from other life sciences programs like the one of NASA, a clear focus in Germany and in Europe, in general, was on the various disciplines in biology rather than on physiology or space medicine. About 50 projects each have been conducted by German scientists in gravitational biology and protein crystallization in microgravity, respectively, roughly 25 in radiation and astrobiology, leaving only 15 projects for space medicine.

Comprehensive overviews on the scientific results of the Shuttle/Spacelab era are provided by NASA for life sciences and for biomedical research, respectively (Naumann et al. 1999; Risin and Stepaniak 2013). Presenting just a few highlights from the German perspective should therefore be sufficient here (Ruyters 2020; see also Chap. 3).

Investigations on gravity responses in microorganisms such as ciliates and Euglena as well as on the alga Chara and higher plants like cress in the 1990s—especially in the timeframe from D-2 (1993) via IML-2 (1994) to STS-81 (1997)—laid the foundation for the high recognition of German scientists in gravitational biology worldwide. Together with their American colleagues, they contributed significantly to the elucidation of the various steps of the transduction chain of the gravity signal in phenomena like gravitaxis and gravitropism. Shuttle/Spacelab experiments also led to breakthroughs in protein crystallization, such as the first successful crystallization of S-layer surface proteins of Archaebacteria, the refinement of structure elucidation of the photosystem I complex—to name only a few examples. In the period between 1988 and 1995, also Ada Yonath, honored with the Nobel Prize in 2009 for her contribution to determining the structure of ribosomes, was involved in 12 space missions. The crystals grown during the Shuttle/Spacelab missions such as D-2, IML-2, and USML-3 were larger and more evenly shaped thus pointing and guiding the way to further experiments on Earth, which then ultimately ended up with the successful elucidation of the ribosome structure and finally with the Nobel prize.

Moreover, German scientists were involved and responsible for measuring the space radiation during many Shuttle/Spacelab missions, due to their long-standing experience and fruitful cooperation with scientists from NASA—a cooperation that is being continued also in the ISS era.

Although relatively few German Shuttle/Spacelab projects were concerned with human physiology aspects, already the first three Spacelab missions FSLP (1983), D-1 (1985), and D-2 (1993) yielded important results being partially in contradiction with common textbook knowledge. For instance, as measured in D-1, central venous pressure of the astronauts decreased inflight, a finding that was so surprising and unexpected that the scientific community would only accept this result after further analysis on D-2 and other missions. In vestibular physiology, the caloric nystagmus was found to occur also in microgravity disproving the theory of the Nobel Prize laureate Barany. Also, scientists began to study the complex interplay of the cardiovascular, respiratory, autonomous, and immune systems thus encouraging a new integrative research approach. This should lead many years later to the detection of a novel mechanism of sodium metabolism (for details, see Chap. 3). Intraocular pressure was found to increase rapidly in the early stages of the flight and then slowly returned to normal levels. These results were supported and detailed later in studies on MIR.

As mentioned already, the successful Spacelab utilization ended in 1998 with the Neurolab mission. After that, the remaining shuttle flights were mainly used for the construction of the International Space Station ISS rather than for international research. The flight of STS-135 finally terminated the Space Shuttle era in August 2011.

2.1.6   MIR: West-European and German Research on the Russian Space Station

Roughly in the same timeframe that saw the development and utilization of Spacelab, cooperation in manned spaceflight also started between Europe and the Soviet Union. Again, Germany and—in this endeavor—also France were the driving forces. After the flight of Sigmund Jähn from the German Democratic Republic in 1978 and the French astronaut Jean-Loup Chrétien in 1982, both utilizing the Salyut-7 Space Station, the newly built MIR station with its first module launched in 1986 was frequently visited by French and German cosmonauts. In addition, ESA

Table 2.3 European cosmonauts on Soviet/Russian Space Stations

Name	Participating organization	Mission	Station	Date
V. Remek (Cze)		Interkosmos	Salyut 6	2.–10.03.1978
S. Jähn (D)	IKF, former GDR	Interkosmos	Salyut 6	26.8.–3.9.1978
J.-L. Chrétien (F)	CNES		Salyut 7	24.6.–2.7.1982
J.-L. Chrétien (F)	CNES	Aragatz	MIR	26.11.–21.12.1988
H. Sharman (GB)	Private financing/UK	Juno	MIR	18.–26.5.1991
F. Viehböck (A)	Austrian Space Agency	Austromir	MIR	2.–10.10.1991
K.-D. Flade (D)	DLR	MIR'92	MIR	17.–25.3.1992
M. Tognini (F)	CNES	Antares	MIR	27.7.–10.8.1992
J.P. Haigneré (F)	CNES	Altair	MIR	1.–22.7.1993
U. Merbold (D)	ESA	Euromir'94	MIR	3.10.–4.11.1994
T. Reiter (D)	ESA	Euromir'95	MIR	3.9.1995–29.2.1996
C. André-Deshays (F)	CNES	Cassiopée	MIR	17.8.–2.9.1996
R. Ewald (D)	DLR	MIR'97	MIR	10.2.–2.3.1997
L. Eyharts (F)	CNES	Pégase	MIR	29.1.–19.2.1998
I. Bella (Slovakia)		Stefánik	MIR	20.–28.2.1999
J.-P. Haigneré (F)	CNES	Perseus	MIR	20.2.–28.8.1999

performed its two Euromir missions in 1994 and 1995. Astronauts from the United Kingdom and Austria in 1991 and from Slovakia in 1999 completed the list of European visitors to MIR (see Table 2.3).

While cooperation with the Soviet Union appeared completely impossible in the early 1980s especially for West Germany, the global political scene had changed dramatically by the end of the decade: Because of glasnost, perestroika, Germany's new "Ostpolitik"—and partially because the D-2 mission had been delayed for several years as a consequence of the Challenger accident—cooperation with the Soviet Union now became possible and even attractive. After all, the Soviet Union and the USA were still the only countries capable of launching crewed space missions. During a state visit by the German chancellor Helmut Kohl to the Soviet head of state, Mikhail Gorbachev, in October 1988, federal minister of research Heinz Riesenhuber signed an agreement providing for the flight of a German cosmonaut to the Soviet/Russian space station MIR. Even in retrospect, it appears almost incredible: Despite all the dramatic political events of those years, including the collapse of the Soviet Union, of the communist regime, and of the Eastern bloc as well as the reunification of Germany, the bilateral MIR'92 mission agreed with the Soviet Union and later conducted by Russia took off punctually on March 17, 1992, from the Baikonur cosmodrome with the German cosmonaut Klaus Dietrich Flade onboard. After this success, German scientists would go on using the Russian

MIR station for nearly a decade particularly for biomedical research. Following two ESA missions, EUROMIR 94 and 95, with the German cosmonauts Ulf Merbold and Thomas Reiter, another German–Russian mission was launched in 1997: MIR 97 with Reinhold Ewald on the crew. In addition, Russian cosmonauts served as test subjects in between these major missions to make the results of medical experiments statistically robust.

In the timeframe from 1988 to 1999, CNES performed six missions to MIR in cooperation with Soviet/Russian partners. All these European missions proved to become especially valuable for European researchers in space medicine since many experiments could be performed on the cosmonauts without having the need for download that was very constrained to 10–15 kg. In fact, a symposium organized by CNES in Lyon in March 2001—a few days before the controlled de-orbiting of MIR—demonstrated the wealth of the new findings obtained not only by the European and Russian scientists but also by the US, Japanese, and other researchers from around the world.

For the German life sciences program, 47 of the 57 German projects dealt with questions relating to space medicine, such as changes in the distribution of fluids along the body axis and the regulation of the cardiovascular and vestibular systems (Ruyters 2008). Because MIR offered cosmonauts a chance to live and work in a microgravity environment for prolonged periods of several months, opportunities for research opened up in some areas of space medicine that had not been possible to tackle during the rather brief shuttle missions. Particularly, these included muscle and bone research and analyses of the psycho-physiological effects of extended stays in space. From preliminary studies during the American Skylab missions in 1973/1974, it was known that sustained weightlessness causes muscles and bones to degenerate. These phenomena resembling age-related or immobilization-induced muscle and bone loss on Earth and obviously occurring in healthy astronauts even in an accelerated manner were now thoroughly addressed by various experiments on the MIR station. Results from these experiments and also from research into the mental and psychological capacity of the cosmonauts on the MIR station have helped to form the basis for planning and preparing ISS and future long-range exploratory missions. The MIR missions also laid the foundations for further successful cooperation between Europe and their Russian partners—especially between ESA and DLR and the Institute of Biomedical Problems (IBMP) in Moscow—in the fruitful utilization of the Russian laboratory of the International Space Station.

2.1.7 The International Space Station ISS: The Biggest Laboratory in Space

At the end of the millennium, both the Spacelab and the MIR era were coming slowly to a close. Fortunately, at the same time, political consensus had been reached

Fig. 2.3 The current configuration of the International Space Station was completed in 2011 consisting of 15 pressurized modules, structural trusses, solar arrays, thermal radiators, and docking ports. The ISS serves as a laboratory for microgravity and space-related research circling the Earth every 90 min at an average altitude of 400 km. (© NASA)

for building the International Space Station ISS as the biggest symbol for international cooperation and technological expertise (Fig. 2.3). After agreements between the USA, Canada, Japan, and ESA were in place, planning and development of ISS elements started worldwide. But as usual, the situation in Europe was complicated. It took more than 10 years to decide on proper European participation, the final decision being taken on the ESA Council at Ministerial Level in 1995. A little later, Russia accepted the invitation of the US to become a partner in the world's largest technological endeavor in January 1998.

Within ESA, Germany again took the leading role with roughly 40% subscription to the respective ESA development and utilization programs. In addition, DLR substantiated its ISS utilization program via bilateral cooperation agreements with all partners, with Russia and NASA being the most important ones. In fact, the first two German ISS experiments in life and physical sciences were already performed in 2001 long before the ESA Columbus module was launched, namely the plasma crystal experiment in cooperation with Russian partners and DOSMAP—measuring the radiation field—in cooperation with NASA.

In February 2008, research activities received a new boost when the European ISS laboratory Columbus was successfully commissioned by the German ESA astronaut Hans Schlegel and went into operation afterwards (Fig. 2.4). By providing lab space and several excellent research facilities especially for life and physical sciences, the bandwidth of life sciences research on the ISS was largely enhanced

Fig. 2.4 Front end of the International Space Station ISS with the shorter European Columbus module provided by ESA and the longer Japanese KIBO module connected to the central US node 2 (Harmony). Both modules house several research racks (insert Columbus) and accommodate external platforms with antennas and scientific payloads. (© ESA, NASA)

(see below and Chap. 3). Before we provide some information on the scientific results obtained so far, a brief summary on the history of ISS, an impressive example of successful international collaboration, seems justified (for further information see Reinke 2007).

2.1.7.1   Brief History of the ISS

For more than two decades, 16 nations are now jointly operating the largest research laboratory ever built in space—the International Space Station (ISS). The five ISS partners are the USA, Russia, ESA, Canada, and Japan. The ISS is not only the greatest science and engineering project in the history of mankind, but also convincing and impressive evidence that international cooperation in the peaceful international utilization of space to the advantage of all partners is both possible and successful. This is true despite numerous delays and technical problems that had to be overcome, including the accident of the space shuttle Columbia in 2003 in the midst of the construction phase of the ISS. Since the first occupants "moved in" on November 2, 2000, astronauts have been permanently living, working, and conducting research on the ISS.

In 1983, the USA had begun to consider building a space station together with partners in Europe, Japan, and Canada. On January 25, 1984, President Reagan commissioned NASA to develop a permanently manned space station that was to be used for scientific and industrial research as well as for the production of special materials and medicines. It was to be launched in 1992, the jubilee year of the rediscovery of America by Christopher Columbus.

In 1985, the ESA Council of Ministers, meeting in Rome, approved Europe's contribution to the American space station. The conditions that would govern the European participation were negotiated during subsequent European-American talks. When the ESA Council of Ministers met next at The Hague in 1987, the European Ministers endorsed the COLUMBUS program. At that time, this program included a module firmly docked to the core station—the later Columbus module—a temporarily manned free-flying laboratory, an unmanned research platform on a polar orbit, and a data relay satellite. However, it quickly turned out that an international space station was a difficult undertaking, for it constituted an entirely novel problem in cooperation within the community of nations. Partner states had to agree not only on a technological concept and the way in which the station was to be used but also on a legal framework. Regulations covering all these questions were encoded in an intergovernmental agreement in 1988.

Only 1 year later, in 1989, the Berlin Wall came down, and a little later, Russia adopted a democratic constitution. As the Cold War between East and West ended, cooperation supplanted competition: In 1993, the USA invited Russia to participate in the ISS program. The other partner countries assented to this proposal in 1994 because it promised many advantages: With Salyut and MIR, Russia had acquired great expertise in designing, building, and managing space stations. Moreover, an additional partner would offer another shoulder to bear the financial burden by, for instance, providing the Soyuz and Proton launchers. As an additional benefit, the Russian space station MIR was used by the ISS partner states in the 1990s to train astronauts in working together in space.

The costs of the new project were growing, as were the technological problems. Together with the worldwide economic crisis in the early 1990s and the participation of Russia in the international space station project, this called for a fundamental reformulation of the concept also for Europe. First, the free-flying European laboratory was canceled, and the docking European laboratory, COLUMBUS, was reduced in size. In 1995, the ESA Council of Ministers at the meeting in Toulouse voted to support the ISS program with additional contributions: in addition to the Columbus lab the development and production of five ATVs (Automated Transfer Vehicle) to be launched by Ariane-5 for safeguarding the transport of supplies and experiment equipment to the station were decided as well as the data relay system for the Russian Zarya module, and the robotic arm ERA for the Russian part of the station. Also, the Columbus Control Center (GSOC) was to be established in Oberpfaffenhofen in southern Germany. By the way, after five successful ATV missions between 2008 and 2014, NASA and ESA agreed that ESA would develop an ATV-derived Service Module as a European contribution to the Orion program for future exploration missions.

On January 29, 1998, representatives of all five ISS partners met in Washington to sign a new intergovernmental agreement so as to provide a basis in international law for the ISS. While it gave more emphasis to the principle of equal partnership than the agreement of 1988, the USA retained its leading role in design and building.

With the launch of the Russian Zarya (dawn) module on November 20, 1998, at Baikonur the construction of the ISS began thus starting the most intense flight activities in the history of space flight. Since November 2000, astronauts and cosmonauts are permanently inhabiting the orbiting station. However, the destruction of the Space Shuttle Columbia as it reentered the atmosphere on February 1, 2003, caused yet another delay of more than 3 years as well as a marked cutback in the number of shuttle flights scheduled before the program was terminated in 2011. Fortunately, shuttle flights could successfully be resumed in 2005 so that the construction of the space station was completed with the launch of the European Columbus module in 2008 and of the six elements of the Japanese module KIBO in 2008 and 2009, thereby adding substantial experiment capabilities for research in life sciences and other scientific domains.

Due to the success of the ISS also regarding the scientific results with great benefit for people on Earth, the ISS partners have extended its lifetime several times. In Europe, where representatives of all member states of ESA meet every 3 years to decide on the ESA programs, funding for the continuation of the ISS and its scientific utilization until 2030 was indicated in the program declaration 2019 and in ESA's long-term plan, which also safeguards further flight of European astronauts (Table 2.4).

2.1.7.2   Benefits of Life Sciences Research on ISS

In the first 5–10 years, ISS utilization only slowly gained momentum, since a big bunch of the crew time was dedicated to setting up, completing, and optimizing vital functions and basic systems. However, scientific utilization dramatically picked up speed with the rapidly increasing availability of laboratory racks and research capabilities. Whereas approximately 30 experiments had been performed during the first crew rotations, this number almost tripled during recent increments. Not only the complexity of the investigations increased but often also their duration, especially in the field of human physiology where proper numbers of subjects' data sets are mandatory. At the same time, the number and the quality of publications grew quickly.

"Astronauts have conducted nearly 3000 science experiments aboard the ISS"— with this headline the well-known journal *Nature* started its news on ISS in November 2020 (www.nature.com/articles/d41586-020-03085-8). Roughly half of them were devoted to life sciences questions—distributed over the ISS partners. Overall benefits to people on Earth—achieved from these experiments—are routinely described to demonstrate the maturity and the impact of the research performed (https://www.nasa.gov/sites/default/files/atoms/files/benefits-for-human ity_third.pdf). The respective websites of the space station partners provide a wealth of information—European experiments supported by ESA are presented in the ERASMUS data archive (www.eea.esa.spaceflight.int). About 200 European life sciences projects have been implemented during these 20 years, covering human

Table 2.4 European astronauts on ISS missions

Name	Participating organization	Mission	Date
U. Guidoni (I)	ESA	ISS flight 6A	19.4.–1.5.2001
C. Haigneré (F)	ESA, CNES	Andromède	21.–31.10.2001
R. Vittori (I)	ESA, ASI	Marco Polo	25.4.–5.5.2002
P. Perrin (F)	NASA, CNES	UF-2	5.–19.6.2002
F. de Winne (B)	ESA	Odissea	30.10.–10.11.2002
P. Duque (E)	ESA	Cervantes	18.–28.10.2003
A. Kuipers (NL)	ESA	Delta	19.–30.4.2004
R. Vittori (I)	ESA, ASI	Eneide	15.–25.4.2005
T. Reiter (D)	ESA/RSA	Astrolab	4.7.–22.12.2006
C. Fuglesang (S)	NASA/ESA	Celsius	9.–22.12.2006
P. Nespoli (I)	NASA/ESA	Esperia	23.10.–7.11.2007
H. Schlegel (D)	NASA/ESA	Columbus	7.–20.2.2008
F. de Winne (B)	NASA/ESA	OasISS	27.5.–1.12.2009
C. Fuglesang (S)	NASA/ESA	Alissé	29.8.–12.9.2009
P. Nespoli (I)	NASA/ESA	MagISStra	15.12.10–24.5.2011
R. Vittori (I)	NASA/ASI/ESA	DAMA	16.5.–1.6.2011
A. Kuipers (NL)	NASA/ESA	PromISSe	21.12.11–30.8.2012
L. Parmitano (I)	ASI/NASA/ESA	Volare	28.5.–11.11.2013
A. Gerst (D)	ESA/NASA	Blue Dot	29.5.–10.11.2014
S. Cristoforetti (I)	ASI/ESA/NASA	Futura	23.11.2014–11.6.2015
A. Mogensen (DK)	ESA/NASA	ArISS	2.–12.9.2015
T. Peake (UK)	ESA/NASA	Principia	15.12.2015–18.6.2016
T. Pesquet (F)	ESA/NASA	Proxima	17.11.2016–2.6.2017
P. Nespoli (I)	ESA/NASA/ASI	Vita	28.7.–14.12.2017
A. Gerst (D)	ESA/NASA	Horizons	6.6.–20.12.2018
L. Parmitano (I)	ESA/NASA	Beyond	20.7.2019–6.2.2020

physiology, biomedical research, gravitational biology, protein crystallization as well as radiation and astrobiology.

Many projects are engaged meanwhile in studying the microgravity-induced challenges of the cardiovascular and the immune system as well as the changes in muscle and bones. In a holistic approach comprising also the influence of stress, nutrition, and salt, scientists study the mechanism of immune deficiency, muscle and bone loss, changes in neurocognitive behavior, as well as the effectiveness of countermeasures. These investigations have provided new knowledge, capabilities, and applications, and have thus contributed greatly to improving health and life on our home planet Earth but also prepare humanity for exploration.

In contrast to the early phase of ISS utilization, more recently benefits are also being generated by private sector players, which make use of ISS to complement their spectrum of laboratory environments for commercial research and for creating innovative new products in the unique conditions not available on Earth. Agencies have implemented processes and allocated ISS resources (i.e., NASA 50%, ESA up to 30%) to foster a broader private sector user community to boost social and innovation impact but also to solidify the foundation of space economy.

After all, ISS utilization—internationally coordinated by the ISS partners and via ISLSWG—has entered, after some delays and problems in the construction phase, meanwhile into a very stable and highly productive period of efficient and successful research promising further breakthroughs to come (see Chaps. 3 and 4). The prolongation of the ISS operation phase until at least 2030 provides additional confidence also to the scientific community and commercial players for ongoing and future utilization and research possibilities in low-Earth orbit.

2.1.8 SIMBOX on Shenzhou: A Model for Sino-German Cooperation in Space?

Roughly at the same time when the Columbus module as major European element was prepared to launch to the ISS, an extraordinary cooperation possibility originated for the German space life sciences program, the SIMBOX/Shenzhou-8 project. It was a peculiar kind of debut when the German SIMBOX incubator housing 17 biomedical experiments took off early in November 2011 as payload of the Chinese Shenzhou-8 spaceship from the Chinese Jiuquan satellite launch center, the Chinese cosmodrome in Inner Mongolia. Fairly soon afterwards, Shenzhou-8 docked with the Chinese space laboratory Tiangong-1, a precursor module of the Chinese space station. Thereby, China became the third nation that had successfully completed docking maneuvers in low-Earth orbit.

The mission was also receiving great international attention because of the SIMBOX project, the world's first bilateral cooperation project with China in manned space flight. For the first time ever, the China Manned Space Engineering Office (CMSEO, now CMSA China Manned Space Office) was cooperating with another nation—Germany—on the basis of an agreement on the peaceful utilization of space for the benefit of both countries (Preu and Braun 2014).

After a mission of 17 days with two successful docking maneuvers with Tiangong-1, the Shenzhou-8 capsule landed on schedule in the desert of Inner Mongolia. Difficult and sometimes tedious discussions and negotiations had eventually brought together different cultures, political systems, ways of thinking, and also representatives from space agencies, industries, and science; and they all achieved their ultimate goal—a successful cooperative science mission.

Altogether 17 experiments had been performed—9 by Chinese scientists, 6 by German scientists, and 2 cooperative projects. One focus of the mission was on gravity perception and signaling pathways in algae, protists, plant cells, and seedlings of higher plants such as Arabidopsis; others were dealing with the effects of microgravity on human neuronal, thyroid cancer cells, and macrophages. One of the joint experiments was a miniature ecosystem designed for investigating fluxes of matter and energy in a closed biological life support system in microgravity, the

Fig. 2.5 First German–Chinese SIMBOX project on a Chinese reentry capsule (Shenzhou-8). The biological samples were processed inside the SIMBOX incubator during the 17-day mission in orbit. After landing of the capsule in the Inner-Mongolian desert, hand-over took place in Beijing for further analyses in the home labs. (© M. Braun, DLR)

other was on protein crystallization with the goal to get new clues for supporting the development of new drugs (Fig. 2.5).

Intensive discussions between Chinese authorities and DLR on future possibilities of cooperation followed. In January 2018, CMSA and DLR signed an agreement on the joint utilization of the Chinese Space Station CSS. The future will show whether and how this chance of further cooperation projects is taken to boost further cooperation projects in space. More recently, also ESA has acknowledged the cooperation possibilities with China. Both partners selected projects from existing pools of experiment candidates to create joint projects to be implemented on ISS and CSS. A first collaborative experiment on space radiation measurements has started on the ISS in March 2020.

2.2 Simulation Studies on the Ground: Preparation for and Complementation of Space Research

Despite the impressive spectrum of microgravity research platforms and flight opportunities available and described above, research opportunities in space remain limited. Therefore, scientists have always looked for suitable simulation options on the ground. Biologists, for example, try to eliminate the effects of gravity on organisms in clinostats. To examine the effects of gravity on humans, medical researchers keep test subjects supine in bed. In isolation studies or studies in extreme locations (for example, at the Antarctic, in submarines), scientists mainly investigate the psychological challenges, which humans also face during space flight. Consequently, simulation studies are a key element of space programs worldwide and yielding insights that are important for astronauts as well as for people on Earth.

2.2.1 Microgravity Simulation and Centrifugation in Gravitational Biology

In gravitational biology, devices to modify the impact of gravity on Earth have a long-standing tradition. Centrifuges are used to increase the gravity level, while devices such as clinostats, random positioning machines, and rotating wall vessels have been developed to produce functional weightlessness or simulated microgravity (Hemmersbach et al. 2018). Since the introduction of the classical clinostat in 1879 by Julius Sachs, various types of clinostats with certain pros and cons have been used such as the 2D, 3D, fast, and slow rotating clinostats. These devices do not abolish the direction of gravity but instead randomize the direction of gravity with respect to the sample over time. In context with space research, Briegleb (1988) has nicely summarized the early results from gravitational biology studies. In the case of magnetic levitation, which was also used by scientists to simulate microgravity, gravity is compensated by creating a counteracting force.

In a recent review, Herranz et al. (2013) provided a comparative study of the most frequently used microgravity simulation devices and illustrate their individual capabilities and limitations. Valuable recommendations are given for a proper selection of adequate devices for defined applications as well as of suitable biological specimen to choose for investigation. Following such recommendations is advisable since there has been a lot of confusion in utilizing the respective methods and in interpreting the results obtained. Due to the limited access to space and the high cost associated, these ground-based studies remain—when applied with caution and expertise—of great importance in preparing and testing the space experiments and for performing well-controlled self-standing experiments on Earth. In most cases however final proof of a new concept or idea can only be achieved by an experiment in a stable and high-quality microgravity environment.

ESA provides access to a whole range of ground-based facilities through its "Continuously Open Research Announcement" opportunity (ESA-CORA-GBF; www.esa.int/Science_Exploration/Human_and_Robotic_Exploration/Research/Bedrest_and_Ground_Studies). The facilities are run by different European institutions such as DLR Cologne (Germany), MEDES Toulouse (France), BIOTESC Lucerne (Switzerland), and ESA/ESTEC in Noordwijk (Netherlands)—to name the most important ones. DLR and CNES provide access to their simulator facilities

Fig. 2.6 Examples of centrifuges and ground-based microgravity simulation devices for gravitational biology research at DLR Cologne. Upper row from left to right: MuSIC—Multiple Sample Incubator Centrifuge with different rotors, fluorescence microscope on a centrifuge, fast-rotating centrifuge microscope. Lower row: Random Positioning Machine (Dutchspace, the Netherlands), slideflask clinostat for cell cultures, 2D clinostat for pipettes, and a submersed 2D clinostat for aquatic samples. (© DLR Cologne)

also via their own national life sciences programs (see also the respective websites) (Fig. 2.6).

2.2.2 Simulation Studies in Integrative Physiology and for Testing Countermeasures

For decades, bed rest and dry or water immersion studies, as well as isolation and confinement studies, have been utilized to mimic the effects of spaceflight on the physiology and psychology of humans. Also, potential countermeasures to mitigate harmful spaceflight effects have been tested (see Corbin and Vega 2019). In addition to the lower costs, perfectly controlled conditions and higher number of test subjects are making these studies ideal for preparing and accompanying spaceflight experiments. They are also most valuable for stand-alone experiments to disclose psychological effects and physiological mechanisms of deconditioning of the human body due to prolonged immobilization and isolation. As for biological projects, also here the final proof of the results obtained requires experiments in space under real microgravity conditions (Table 2.5).

Table 2.5 Analogs for space-related human health and performance studies

Study type	Major locations
Bed rest studies	Envihab (DLR Cologne), MEDES Space Clinic (Toulouse), Planica Nordic Center (Slowenia), several in the USA
Dry/water immersion	IBMP Moscow, MEDES Space Clinic (Toulouse)
Isolation and Confinement	SIRIUS/NEK (Nazemnyy experimental'nyy kompleks) at IBMP (Russia), HERA, Houston (USA), CELSS, Shenzhen (China)
Antarctica Stations (isolation)	Concordia (ESA), Neumayer III (AWI/DLR), McMurdo (NASA)

2.2.2.1   Bed Rest Studies

Of course, the history of bed rest started long before human missions into space. Before the nineteenth century, bed rest as a means to treat diseases was usually avoided, because periods of prolonged bed rest often led to economic problems such as the loss of employment. This changed in the nineteenth century when bed rest was first introduced as a medical treatment. At that time, any adverse consequences were attributed to the disease rather than to the bed rest treatment as such. Only after a few decades, physicians have become aware that bed rest is not the cure for everything and might even be the root cause for further health problems. Respective early observations by Whedon and coworkers (Deitrick et al. 1948) soon found application in space research and prompted space agencies to start thinking about the performance of specific bed rest studies. This seemed the more necessary since obviously the negative consequences of spaceflight on the astronauts' health and performance were not only due to their relative inactivity within the spacecrafts—as assumed in the early days—but even more due to microgravity affecting blood circulation, fluid, and electrolyte balance, causing postural and vestibular problems as well as changes in the immune, muscle, and bone systems.

After some studies that used chair-sitting for a few days, placing healthy volunteers in bed thus became the model of choice for inducing and studying the effects of prolonged spaceflight and for testing potential countermeasures. Results of the early studies have nicely been summarized by Sandler and Vernikos (1986), those performed in the timeframe 1986–2006 by Pavy-Le Traon et al. (2007). A more recent compilation of bed rest studies has been published by Goswami et al. (2019). Further information can be found in the respective chapters of several books of our SpringerBriefs series (Blottner and Salanova 2015; Heer et al. 2015; Gunga et al. 2016).

As evident from these publications, especially NASA and Europe took the lead in developing standard protocols for bed rest studies and established centers for running bed rest studies in the US as well as in France (MEDES Toulouse) and Germany (DLR Cologne) and in the future also in Planica (Slovenia). In most cases, bed rest studies were organized in international settings based on cooperation agreements and standard protocols established in the frame of ISLSWG, the International Space Life Sciences Working Group. Protocols and research projects were

Table 2.6 Overview of European bed rest studies between 2000 and 2020

Name	Year	Location	Duration (days)	Countermeasure used
LTBR	2001/ 2002	MEDES, Toulouse (FR)	90	Flywheel exercise, bisphosphonate
STBR	2001/ 2002	DLR, Cologne (DE)	14	Caloric variations in nutrition, amino acid infusion
BBR-1	2003/ 2004	Charité Berlin (DE)	56	Vibration exercise
WISE	2005	MEDES, Toulouse (FR)	60	Combined resistive exercise, aerobic exercise, lower body negative pressure, nutritional supplement
BBR-2	2007/ 2008	Charité Berlin (DE)	60	Vibration exercise
BRAG-1	2010	MEDES, Toulouse (FR)	5	Artificial gravity
NUC	2010	DLR, Cologne (DE)	21	Nutritional supplement
BRAG-2	2010/ 2011	DLR, Cologne (DE)	5	Artificial gravity
MEP	2011/ 2012	DLR, Cologne (DE)	21	Nutritional supplement
MNX	2012/ 2013	MEDES, Toulouse (FR)	21	Nutritional supplement combined with resistive exercise
RSL	2014/ 2015	DLR, Cologne (DE)	60	Reactive jumps
Cocktail	2017	MEDES, Toulouse (FR)	60	Nutritional supplement
AGBRESA	2019	DLR, Cologne (DE)	60	Artificial gravity

optimized and integrated by asking the scientists to share the test subjects as well as the data for better interpretation of their findings. In the first two decades of the twenty-first century, more than ten bed rest studies have been conducted following these standard rules—more and more directed towards preparing long-duration exploration missions (Table 2.6).

In this context, especially the value of testing and validating of countermeasures against the negative effects of long-duration bed rest or microgravity must be emphasized. Several methods and devices have been investigated in bed rest studies providing valuable information on their efficacy for the different physiological systems in humans. Currently, vibration exercise and reactive jumps on a newly developed sledge jump system seem to be most promising (for details see Sects. 4. 7.2 and 4.7.3). These achievements are, of course, also of increasing benefit to maintaining health and performance for humans on Earth. Certainly, also the generally shorter stay of patients in hospitals applied today is not the least due to the results of bed rest studies performed in the context of human spaceflight.

In Russia, water and especially dry immersion has been extensively used in addition to bed rest—the advantage being that also the transverse gravity vector Gz is reduced by this method (for review see Tomilovskaya et al. 2019). However, the acceptance by the test subjects especially for longer periods is difficult to achieve due to the inconveniences associated. Nevertheless, a series of dry immersion studies is planned by ESA for upcoming years at MEDES.

2.2.2.2 Isolation and Confinement Studies

While in bed rest studies the focus of research is on analyzing the consequences of this setting on the various physiological systems rather than on psychological issues, isolation and confinement studies are specifically organized to analyze the performance of the subjects and aspects of crew interaction. Initiated already in the 1980s, isolation and confinement studies have come into a new focus in context with the recent international exploration strategies as well as with the current COVID-19 pandemic. In this context, Choukèr and Stahn (2020) have summarized just recently the history and important results of these studies emphasizing the value of shared learnings from spaceflight and what they call "the largest isolation study in history—COVID-19" (see also Sect. 3.5.6).

The settings can vary—ranging from closed chambers such as successfully applied for the famous Mars500 study at IBMP in Moscow, in HERA and SIRIUS to remote and confined places on Earth such as Antarctica stations. In HERA (Human Exploration Research Analog) NASA annually sponsors campaigns of 30–45 days of isolation with a crew of four. Five campaigns have been completed to date with the participation of several German research studies by DLR supported scientists. Isolation studies of extended length ranging from 4 months up to 1 year are being conducted at the NEK (Nazemnyy experimental'nyy kompleks) facility at IBMP Moscow. These studies called SIRIUS (Scientific International Research in Unique Terrestrial Station) are coordinated and conducted in international cooperation, which is—by the way—a consistent attribute of most of these activities including bed rest studies.

Space agencies in Europe have become especially active in Antarctica in recent years. Living in an Antarctica station is remarkably close to that on the surface of another planet. The crew is even more cut off from civilization and resources than astronauts on the ISS, making this environment highly suitable to study the effects of isolation and confinement, and, thus, for preparing upcoming exploration missions to Moon or Mars. In fact, since 2005 ESA organizes isolation studies at the Concordia station via agreements with the French IPEV Institute and the Italian PNRA Antarctic Program. Investigations on changes of the immune system, on alterations in circadian rhythms and body temperature regulation as well as on coping with stress are in the focus of the research projects selected in competition among European scientists.

In addition to their participation via the ESA program, German scientists are also active at the German Neumayer station via a bilateral cooperation agreement concluded in 2009 between DLR and the Alfred-Wegener-Institute running the station. Studies on the effects of isolation and confinement on the cardiovascular system, on sleep and performance, on stress and the immune system as well as on hormones and vitamin D are in the focus and have led to interesting results—of importance not only for astronauts and the winter-over subjects in Antarctica but also of benefit generally for humans in extreme environments (Gunga 2015).

In summary, bed rest, as well as isolation and confinement studies usually performed in international cooperation and under controlled and standardized conditions, yield important results on the challenges of these conditions resembling spaceflight with respect to the physiology and psychology of humans as well as on the effectiveness of countermeasures. These results are not only of utmost importance for preparing long-term exploratory space missions but also of great benefit for humans on Earth especially in the aging societies of the industrialized countries. More is certainly to be expected from the currently ongoing and future studies.

References

Blottner D, Salanova M (2015) The neuromuscular system: from earth to space life science, SpringerBriefs in space life sciences. Springer, Cham. https://doi.org/10.1007/978-3-319-12298-4

Briegleb W (1988) Ground-borne methods and results in gravitational cell biology. Physiologist 31:44–47

Buckey JC, Homick JL (2003) The Neurolab spacelab mission: neuroscience research in space. NASA SP-2003-535

Choukèr A, Stahn AC (2020) COVID-19 – the largest isolation study in history: the value of shared learnings from spaceflight analogs. Npj Microgravity 6:32. https://doi.org/10.1038/s41526-020-00122-8

Corbin B, Vega LM (2019) NASA's use of ground and flight analogs in reducing human risks for exploration. Front Physiol. Conference abstract: 39th ISGP meeting & ESA life sciences meeting. https://doi.org/10.3389/conf.fphys.2018.26.00044

Deitrick JE, Whedon GD, Shorr E (1948) Effects of immobilization upon various metabolic and physiologic functions of normal men. Am J Med 4:3–36

Franke B (2007) Forschungsraketen (in German). Stedinger Verlag, Lemwerder

Goswami N, van Loon JJWA, Roessler A, Blaber AP, White O (2019) Gravitational physiology, aging, and medicine. Front Physiol 10. https://doi.org/10.3389/fphys.2019.01338

Gunga HC (2015) Human physiology in extreme environments. Elsevier, Amsterdam

Gunga HC, Weller von Ahlefeld V, Appell Coriolano HJ, Werner A, Hoffmann U (2016) Cardio-vascular system, red blood cells, and oxygen transport in microgravity, SpringerBriefs in space life sciences. Springer, Cham. https://doi.org/10.1007/978-3-319-33226-0

Heer M, Titze J, Smith SM, Baecker N (2015) Nutrition, physiology and metabolism in spaceflight and analog studies, SpringerBriefs in space life sciences. Springer, Cham. https://doi.org/10.1007/978-3-319-18521-7

Hemmersbach R, Häder DP, Braun M (2018) Methods for gravitational biology research. In: Gravitational biology I – Gravity sensing and graviorientation in microorganisms and plants, SpringerBriefs in space life sciences. Springer, Cham. https://doi.org/10.1007/978-3-319-93894-3

Herranz R, Anken R, Boonstra J, Braun M, Christianen PCM, de Geest M, Hauslage J, Hilbig R, Hill RJA, Lebert M, Medina FJ, Vagt N, Ullrich O, van Loon JJWA, Hemmersbach R (2013) Ground-based facilities for simulation of microgravity: organism-specific recommendations for their use, and recommended terminology. Astrobiology 13:1–17

Launius R (2020) Highlights of human spaceflight: the United States. In: Young LR, Sutton JP (eds) Handbook of bioastronautics. Springer, Cham. https://doi.org/10.1007/978-3-319-10152-1_79-2

Naumann RJ, Murphy KL, Lewis ML (1999) Spacelab sciences results study, Microgravity life sciences, vol III. NASA Study NAS8-97095

Pavy-Le Traon A, Heer M, Narici MV, Rittweger J, Vernikos J (2007) From space to earth: advances in human physiology from 20 years of bed rest studies (1986–2006). Eur J Appl Physiol 101:143–194

Pletser V, Winter J, Duclos F, Bret-Dibat T, Friedrich U, Clervoy JF, Gharib T, Gai F, Minster O, Sundblad P (2012) The first joint European partial-g parabolic flight campaign at moon and mars gravity levels for science and exploration. Microgr Sci Technol 24:383–395

Pletser V, Rouquette S, Friedrich U, Clervoy JF, Gharib T, Gai F, Mora C (2015) European parabolic flight campaigns with Airbus ZERO-G: looking back at the A300 and looking forward to the A310. Adv Space Res 56:1003–1013

Preu P, Braun M (2014) German SIMBOX on Chinese mission Shenzhou-8: Europe's first bilateral cooperation utilizing China's Shenzhou programme. Acta Astron 94(2):584–591

Reinke N (2007) The history of German space policy - ideas, influences, and interdependence 1923–2002. Editions Beauchesne, Paris

Risin D, Stepaniak PC (2013) Biomedical results of the space shuttle program. NASA SP-2013-607

Ruyters G (2008) Raumfahrtmedizin in der Bundesrepublik Deutschland. In: Wirth F, Harsch V (eds) 60 Jahre Luft- und Raumfahrtmedizin in Deutschland nach 1945. Rethra Verlag, Neu Brandenburg, pp 91–122

Ruyters G (2020) Highlights of human spaceflight in Europe: ESA and DLR. In: Young LR, Sutton JP (eds) Handbook of bioastronautics. Springer, Cham. https://doi.org/10.1007/978-3-319-10152-1_70-1

Ruyters G, Friedrich U (2006) From the Bremen Drop Tower to the International Space Station ISS – ways to weightlessness in the German space life sciences program. Signal Transduct 6:397–405

Ruyters G, Preu P (2010) Vom Bremer Fallturm zur Internationalen Raumstation – Faszinierende Forschung in Schwerelosigkeit. In: Schwerelos – Europa forscht im Weltraum. Spektrum der Wissenschaft Extra. Ort, Heidelberg, pp 14–19

Sandler H, Vernikos J (eds) (1986) Inactivity: physiological effects. Academic, New York

Schmidt W (2010) Gravity induced absorption changes in Phycomyces blakesleanus and coleoptiles of Zea mays as measured on the drop tower in Bremen (FRG). Microgravity Sci Technol 22:79–85

Schmidt W, Galland P (2004) Optospectroscopic detection of primary reactions associated with the graviperception of Phycomyces: effects of micro- and hypergravity. Plant Physiol 135:183–192

Tomilovskaya E, Shigueva T, Sayenko D, Rukavishnikov I, Kozlovskaya I (2019) Dry immersion
    as a ground-based model of microgravity physiological effects. Front Physiol. https://doi.org/
    10.3389/fphys.2019.00284

Chapter 3
Success Stories: Incremental Progress and Scientific Breakthroughs in Life Science Research

Abstract Research in space life sciences is a multidisciplinary activity that has—over several decades—resulted in many incremental new findings, extensions to our knowledge, and considerable technological progress rather than in just a few spectacular breakthroughs. Although the latter are easier to be appreciated by the public and by politicians, science and research also in terrestrial settings usually take time, money, and patience before a significant progress becomes visible. Nevertheless, we will try in this chapter to convince the dear reader that space life sciences have generated numerous success stories over the years, be it in the form of incremental progress or as real breakthroughs. An even broader view on the accomplishments that are currently delivered by the research and development activities specifically on the International Space Station is published frequently by the ISS partners in "ISS Benefits for Humanity" (latest issue 2019; https://www.nasa.gov/sites/default/files/atoms/files/benefits-for-humanity_third.pdf).

Keywords Breakthroughs · Earth benefits · Gravitational biology · Integrated human physiology · International Space Station · Space life sciences · Space medicine

3.1 Introduction: From the Impact of Gravity on Molecules and Cells to the Human Body as an Integrated Physiological System

For 4 billion years, life on our planet has been inextricably interwoven with the force of gravity as a key factor in evolution. Life on Earth is subject to the influence of gravity; in most cases, its significance for many vital functions can only be explored in its absence or by altering gravitational forces. From experiments in microgravity, scientists learn about the mechanisms by which all living beings, plants and animals, protozoa, as well as humans, perceive and respond to gravity with special emphasis on the mechanisms of gravity-mediated signaling. Such insights are of great

© Springer Nature Switzerland AG 2021
G. Ruyters et al., *Breakthroughs in Space Life Science Research*, SpringerBriefs in Space Life Sciences, https://doi.org/10.1007/978-3-030-74022-1_3

importance in basic research for they help us to improve our understanding of the origin, spreading, and evolution of life on our home planet.

Furthermore, investigations in microgravity have demonstrated their eminent importance in application-oriented research such as in biotechnology and medicine. They provide us with knowledge about the interaction between the various systems of the human body such as muscles and bones, the heart, the cardiovascular system, and the immune system thus stimulating a new perspective, namely, to view the human body as an integrated system. This knowledge, in turn, is incorporated in the diagnosis and therapy of the sick. Indeed, research under space conditions has brought forth innovative therapies as well as new instruments for the diagnostics, treatment, and training for maintaining health and performance in humans. Both objectives—researching basic vital functions and developing new methods of diagnosis and therapy in medicine—will also play a crucial role when humankind and space agencies prepare the ground for future long-range missions to the moon or elsewhere. In the following, we will first summarize some of the scientific key findings in biology and human physiology before we will focus on innovative developments especially for biomedical diagnostics and human health maintenance in Chap. 4.

3.2 Key Findings and Breakthroughs in Gravitational Biology

Gravity is the only critical factor in the growth processes of plants providing a constant environmental cue for orientation and exploiting the space below, on, and above the surface of the Earth. Roots grow towards the Earth's center to anchor the plant in the soil and supply water and nutrient salts. Shoots and stems grow away from the Earth's center to expose the plant's leaves to sunlight thereby securing optimum photosynthesis. This gravity-directed growth behavior of plant organs is called gravitropism. Also, numerous aquatic single-cell organisms like, e.g., Euglena or Paramecium use gravity for orientation and can thus choose the place offering the best possible conditions for their growth and survival. This behavior is called gravitaxis. Given the great significance of these processes and the key role of gravity, scientists have tried to unravel the mystery of gravity perception, gravity transduction, and gravity effects on plant morphology and physiology for nearly 150 years, when naturalists like Darwin, Pfeffer, and Sachs were already fascinated by the gravity response of plants. As demonstrated below, microgravity research has significantly contributed to the considerable progress achieved in this scientific endeavor over the past few decades. More detailed overviews can be found in recent volumes of our SpringerBriefs Series on "Space Life Sciences" (Braun et al. 2018; Hanke et al. 2018).

3.2.1 Gravitaxis in Euglena: Space Research Helps to Settle a 100-Year Argument—No, It Is Not a Buoyancy Effect—It Is Active Gravity-Sensing

Many microorganisms that live in water use gravity for their orientation in search of the optimal ecologic environment. For many decades, there has been disagreement among scientists as to whether the response of these organisms to gravity is a result of simple physics, or whether any physiological processes are involved. Experiments on parabolic flights and TEXUS sounding rockets have provided unambiguous prove that microorganisms possess complex physiological mechanisms for the perception of gravity (Häder et al. 2005; Braun et al. 2018). Evolution has produced at least two different systems: One is to be found in the flagellate Loxodes, which has acquired a special organelle—called the Mueller vesicle (Müllersches Körperchen)—with which it perceives gravity by registering the deflection of the mass-holding structure under the gravitational load. Euglena and Paramecium, by contrast, recognize the direction of the gravitational pull from the difference in densities between the cell fluid and the surrounding medium.

Thanks to experiments in microgravity, the individual steps of the gravity-directed response process, i.e., the perception of the original stimulus, its transformation into a signal within the cell, and the eventual physical response mechanism are now no longer a secret. Passing through a specially located mechanosensitive channel in the cell membrane, calcium enters the cell every time the cell's orientation in space changes in a certain way. Once inside the cell, calcium passes through a series of proteins such as calmodulin and activates several enzymes. One of these biological catalysts converts adenosine triphosphate (ATP)—the cell's "energy currency"—into cyclic adenosine monophosphate (cAMP). Acting as a second messenger, this molecule is abundantly present in cells and regulates protein synthesis and the formation of ion channels. However, the exact way in which cAMP influences flagellar motility and thus ensures the organism's preferred orientation in the water column is still unknown. Recent findings suggest that the phosphorylation of proteins, a widely observed cellular mechanism that controls protein activity, is influenced by changed gravity conditions, too. A model of the basically complete gravitaxis sensory transduction chain in Euglena has been published recently (Prasad et al. 2020) representing the result of about three decades of research in space and on ground (Fig. 3.1). In addition, this research is an excellent example for patiently and effectively making use of all proper microgravity platforms provided by the European and German space programs that are required to tackle the challenges of the next step in disclosing the complete signal transduction chain.

Incidentally, the image processing system ECOTOX that had been developed by scientists of Erlangen University to study the motional behavior of Euglena has meanwhile been introduced in many countries as a device for monitoring wastewater samples in the context of freshwater quality control. This is because a change in motional behavior of Euglena has been identified to be a powerful indicator of

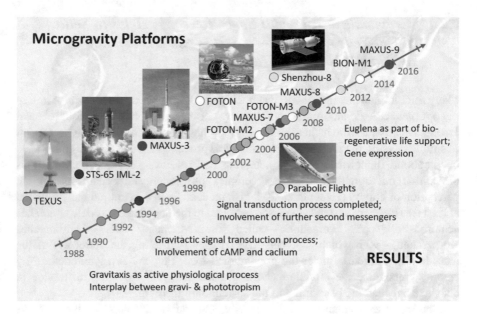

Fig. 3.1 In a step-by-step approach and by utilizing the various microgravity platforms available through DLR and ESA, the gravitactic signaling pathway of Euglena has been disclosed. (© M. Braun)

increased pollutant concentrations (Ahmed and Häder 2011; Häder and Erzinger 2015).

3.2.2 Where to Grow: Plant Gravity-Sensing and Graviorientation

Great scientific progress and new textbook knowledge have also been accomplished in plant gravitropism. It is the mechanism plants use not only for properly orienting roots and shoots independent on highly variable other environmental cues but also to precisely align almost all organs in a specific angle with respect to the gravity vector that is most beneficial for exploring and exploiting resources on, below, and above the surface of the Earth.

Plant biology research in microgravity—over the past decades—has demonstrated that plants can grow seed-to-seed in space and can be cultivated successfully in extraterrestrial habitats (for review see Wheeler 2017). In addition, microgravity experiments have considerably contributed to our understanding of the cellular and molecular mechanisms of gravity-sensing and signal transduction in plants (Ruyters and Braun 2014; see also https://www.nasa.gov/mission_pages/station/research/benefits).

Single-celled, tip-growing rhizoids and protonemata of the green alga Chara have been established as model systems for gravitropism (Braun et al. 2018). The gravitropic signaling pathway is rather short, limited to a small region at the tip of these cells, which are easily accessible by numerous molecular, physiological, and microscopical methods (Braun and Limbach 2006). Experiments on TEXUS rockets and on Space Shuttle missions have delivered clear evidence for the active role of the force-generating actomyosin system in the gravity-sensing process. Gravity sensing relies on statoliths ($BaSO_4$-crystal-filled vacuoles), which are kept in a dynamically stable position above the cell tip by actomyosin forces that precisely counteract the gravitropic pull on the statoliths. Even the slightest deviation of the cells from a vertical orientation leads to gravity-driven sedimentation of the statoliths towards the lower cell flank. However, this statoliths displacement is affected by actomyosin forces. They direct the statoliths towards gravireceptor proteins located in the subapical region of the plasma membrane. Once activated, the receptor generates a physiological signal that intervenes in the growth process. By affecting calcium channels, a local reduction of exocytosis on the lower cell flank causes asymmetric growth of the opposite cell flanks. This triggers gravitropic bending of the cells and the return of the plant organ to its original vertical direction of growth. Results of several parabolic flights have indicated that also in higher-order plants membrane-bound gravity receptors are activated by directly interacting with proteins on the statoliths surface as was clearly shown in Chara rhizoids and protonemata (Braun et al. 2018).

In gravity-sensing cells of higher plants, called statocytes, the mechanism by which the gravity-directed movement of amyloplast-statoliths triggers the gravitropic readjustment of the plant organ (roots, shoots, branches, etc.) is more complex. It also involves the actomyosin cytoskeleton system, but rather as a damping system modifying the sedimentation process more in a way to prevent too fast responses to only short and transient gravitropic stimulation, i.e., like corn stalks swaying in the wind. It furthermore involves a signaling cascade, the modulated flux of the plant hormone auxin under the control of secondary messengers, several associated proteins, channel and carrier proteins (for review see Su et al. 2017; Braun et al. 2018; Kiss et al. 2019; Muthert et al. 2020). Thanks to advanced methods in molecular and cell biology also available on ISS, researchers have gathered information on the molecular basis of gravity sensing and orientation in higher plants (for review see Hanke et al. 2018; Braun et al. 2018).

In addition, only experiments in microgravity could elucidate the interaction between light- and gravity-induced pathways and growth responses in plant organs (Vandenbrink et al. 2016) which have not been detected on Earth due to the dominance of the gravity-mediated responses. This is an impressive example for the usefulness of the microgravity environment as a tool for unraveling sensitive biological pathways, which—in the gravity environment on Earth—lie hidden and are not accessible. In the absence of gravity as dominating environmental cue for orientation however plant growth is guided by other tropisms like phototropism, hydrotropism, and thigmotropism, which need to be understood to prepare proper technologies for farming in space (for a recent review see Muthert et al. 2020).

Experiments on parabolic flights, sounding rockets, satellites, and in the Bremen drop tower with cultures of plant cells, mostly from Arabidopsis, have also contributed greatly to the major progress in our understanding of genetic, biochemical, and metabolic changes triggered by gravity (for review see Hanke et al. 2018). The results clearly indicate that even in cells, which are not specialized for gravity-sensing, gravity effects and corresponding metabolic changes occur in the process of adapting to the novel space environment. Scientists were able to demonstrate by means of molecular analyses that reducing the gravity level causes an increased influx of calcium ions, which, as a result of enzyme activation (NADPH oxidase), leads to increased production of hydrogen peroxide (H_2O_2). Both these messengers intervene in the regulation of gene and protein expression and can, via protein activation (protein phosphorylation), modify the biological function of a cell. These results bring scientists one step closer towards solving the mystery of how cells respond and adapt to external stimuli. Experiments carried out by a science team from the University of Freiburg with Arabidopsis seedlings point in the same direction. They were able to show that changed gravitational conditions massively reprogram cellular regulatory networks. One key mechanism involved appears to be the transport of the growth hormone auxin encoded by the AtPIN3 gene (Braun et al. 2018; Hanke et al. 2018).

Insights into cellular and molecular responses of plants adapting to the gravity stimulus-free environment of space rapidly increased with the completion of the ISS as a permanent microgravity research platform in low-Earth orbit. In only two decades, a wealth of information has been gathered by analyzing plants which have been grown on ISS. Gravity-related effects have been found on phototropism (Vandenbrink et al. 2016 and references therein), the plant cell cycle (Matia et al. 2010), on the composition and characteristics of the cell wall (Johnson et al. 2017), on the molecular and gene expression level (spaceflight transcriptome; Paul et al. 2017 and references therein).

Thus, utilizing the ISS and other flight opportunities—complemented by the respective ground research—has significantly helped to disclose gravity effects in plants and the mysteries of plant gravity sensing and orientation. The ISS also provides the perfect research platform for exploiting the capabilities of plants needed to support exploratory missions to the Moon and Mars and living in space.

3.2.3   Gravity Sensing in Animals

Like many protozoa and plants, most if not all invertebrates and vertebrates have evolved sensory structures to perceive the gravity vector for orientation and postural control. The unique stimulus-free microgravity environment of space has frequently been used as a tool to dissect gravity-sensing mechanisms and to address new hypotheses (Boyle and Hughes-Fulford 2020). We are limiting our report on spaceflight-related results in this field of research to the following questions: Are functional sensory systems developing in the absence of gravity or—more general—

under altered gravitational conditions? Are sensory structures and functions adapting to new gravitational environments?

Before answering these questions, a few remarks should help to explain, why we refrain from going into more details in our book concerning animal experimentation in space. Historically, the Soviet Union and the United States launched several suborbital and orbital missions with mice, rabbits, dogs, and primates before they sent Yuri Gagarin and John Glenn into space. Later, flights of animals continued for basic research purposes in parallel with human spaceflight. Procedures can be perfected on animals, larger numbers can be investigated providing sample sizes for greater statistical power than is possible with human subjects alone, and usually the animal life cycle is much shorter enabling to analyze processes such as muscle and bone loss in time-lapse. Results of space shuttle missions, especially between 1983 and 2003, from the Russian BION M-1 mission in 2013 as well as from the ISS, have been summarized recently (Risin and Stepaniak 2013; Andreev-Andrievskiy et al. 2014; Sandonà et al. 2012; Ronca et al. 2019). All in all, these experiments provided a deeper understanding of the microgravity effects on various physiological systems. However, despite these interesting results and other advantages, animal experimentation especially with mammals has not been a priority in European space life sciences programs over decades.

Therefore, scientists in Europe mainly focused on smaller vertebrates, especially fish, for their space studies on development and gravity-sensing. In fact, it is safe to say that—based on the vast amount of data available from this research—the answer to the above questions on gravisensing in microgravity is positive: Although all sensory mechanisms, undoubtedly, have perfectly adapted to the Earth's gravitational field, several developmental studies in microgravity showed that gravity is not a prerequisite for the development of functional sensory systems (see, e.g., Wiederhold et al. 1997), and it certainly was an exciting moment, when in 1961, Yuri Gagarin confirmed that his senses were still functional in the microgravity conditions of space. However, adaptation to the new gravitational environment certainly takes place. The sensory systems require inertial masses, called otoliths, statoliths, or otoconia, resting on mechanosensory epithelia. The size of otoconia was found to be decreased in hypergravity and increased in microgravity in developing but not in mature vertebrates and invertebrate species (for review see Jamon 2014; Hilbig and Anken 2017). The growth of otoliths is regulated via a negative feedback loop that involves otolithic calcium carbonate deposition probably under control of a network of regulatory genes. Altered gravity also affects the sensory epithelium; e.g., the growth of vestibular neurons was found to be decreased and cerebellar endings in the vestibular nuclei were underdeveloped (Demêmes et al. 2001). Thus, at least during a critical development phase altered gravity conditions affect the plasticity of the otoliths, the sensory epithelium, and the projections from the gravireceptor to the brain and spinal cord (Jamon 2014)—adaptational modifications that may lead to altered postural control and spatial orientation.

3.2.4 Bioregenerative Life Support Systems: A Prerequisite for Long-Term Exploration of Moon and Mars

For nearly 60 years, humans have been living and working in space. The harsh conditions prevailing there—vacuum, extreme temperatures, high radiation exposure, microgravity—have made it necessary to create special habitats for the astronauts, i.e., spacecrafts such as the Spacelab/Space Shuttle system and various space stations from Skylab, Salyut, and MIR to the ISS. Within such an artificial habitat, so-called life support systems provide fresh air, regenerate condense and waste water, and eliminate harmful gases.

Starting with rather simple systems on the first manned space capsules, physicochemical systems have evolved into today's highly complex and reliable advanced life support systems on the ISS that have proven their value to perform all aspects of the abovementioned tasks. However, for long-term missions to distant planets or for Moon or Mars bases, this concept is not realistic because the astronauts will be completely cut off from resupplies on their journey. Therefore, step-by-step, current physicochemical systems must be complemented and even partially replaced by biological systems to achieve complete closure of the systems at the end. In addition, the food needs to be supplied as well. These considerations have stimulated scientists and engineers for decades to develop and test bioregenerative life support systems (BLSS). On Earth, major activities were performed in Russia (BIOS project), in the US (Biosphere-2), in Japan (Closed Ecological Experiment Facility CEEF), and—more recently—in China (Lunar Palace 1). Valuable overviews on these and some other terrestrial activities have been published by Kern (2009) and more recently by Chinese scientists (Guo et al. 2017) as well as by Schubert in his Ph.D. thesis (Schubert 2018).

However, at the end, all these terrestrial BLSS could only partially ensure proper CO_2 removal, O_2 production, and sufficient food supply. This demonstrates that the goal to establish a fully functional bioregenerative and largely closed life support system is an extremely ambitious and tedious one and is most likely only achievable in a step-by-step approach. This is especially true for the planned utilization in spacecrafts, where microgravity leads to additional complications.

3.2.4.1 The ESA MELISSA Project

A stepwise approach is being followed by ESA with its project MELISSA (Micro-Ecological Life Support System Alternative). Established some 30 years ago in 1989, the MELISSA project was formed to create a closed-loop artificial ecosystem for long-duration spaceflights, mainly based on aquatic systems. The MELISSA consortium is a partnership of independent organizations such as universities, research centers, SMEs (small and medium enterprises), and bigger industrial partners such as THALES Alenia. Today, more than 40 partners—especially from Spain, Belgium, France, and Italy (altogether 13 countries)—make up the MELISSA community. Despite these large efforts, progress is judged as rather slow and realization in the form of a ready-to-use functional system on Moon or Mars is far away (for more details see www.esa.int/Our_Activities/Space_Engineering_Technology/Melissa).

3.2.4.2 Bioregenerative Life Support in the German Life Sciences Program

BLSS have a long tradition within the German Space Life Sciences program with first activities starting in 1986 (for review see Häder et al. 2018). From the very beginning, the focus in Germany has been on aquatic systems, since fish and aquatic plants are easier to kccp than land animals—not least because of the waste problems. The philosophy behind the different projects remained basically the same over the years: Before biological life support systems can be used in space, scientists need to understand how the organisms involved interact. Therefore, an extensive program covering aspects of physiology, morphology, ecology, genetics, and developmental biology was already set up between the late 1980s and the early 1990s.

CEBAS (1986–2003) The Closed Equilibrated Biological Aquatic System (CEBAS) was a largely closed aquatic system inhabited by fish, snails, hornwort (*Ceratophyllum demersum*), and microorganisms. The main function of the hornwort was to produce oxygen for the fish and snails and process the carbon dioxide exhaled by them, while the microorganisms were mainly concerned with converting ammonium ions into nitrite and nitrate. Via an agreement with NASA, two successive flights for the facility called CEBAS mini-module on the space shuttle were performed: STS-89 in January 1998 and STS-90 Neurolab in April 1998. Unfortunately, the third shuttle flight of the CEBAS mini-module on STS-107 ended tragically with the loss of the Columbia shuttle during its descent on February 1, 2003. Despite the Columbia accident, CEBAS was an extraordinarily successful research and development project.

The inflight experiments and the supporting studies on the ground proved that the development of the CEBAS mini-module had produced a stable system capable of supporting an extensive research program and constituting an aquatic, largely closed bioregenerative life support system. The fish behaved and developed normally in weightlessness; the development of their embryos remained unaffected. The snails similarly displayed no behavioral changes. Photosynthesis activity, as well as other physiological parameters of the aquatic plants, did not differ from those of the controls on the ground so that growth was normal (for further details see Blüm 2003).

Aquacells/OmegaHab (2000–2013) After the Columbia space shuttle accident, the entire program had to be reoriented. Also, greater emphasis was given to the botanical element. Euglena, a single-celled alga with a good photosynthesis performance, was chosen as the botanical component.

Fig. 3.2 The OmegaHab Bioregenerative Life Support Box is deinstalled from the Russian BION-M1 capsule after landing in Kazakhstan. (© M. Braun)

Preliminary experiments to study gravitaxis and photosynthesis efficiency were conducted in the Aquacells device on Russian FOTON M-2 satellite in June 2003 by scientists from Erlangen University. On the FOTON M-3 mission in September 2007, the algae reactor was combined with a fish tank. Called OmegaHab, this system subsequently went on the BION M-1 satellite mission in May 2013. A whole spectrum of organisms was involved such as Euglena, hornwort (*Ceratophyllum demersum*), an aquatic plant, tilapia larvae (*Oreochromis mossambicus*), *Hyalella azteca*, a Mexican freshwater crustacean, and a few ram's horn snails (*Biomphalaria glabrata*).

The system worked perfectly until problems with the LED light devices occurred. Despite this technical failure, OmegaHab can be rated as a successful project. The experiments on FOTON M-3 and BION M-1 provided valuable data on how cells, organs, and entire organisms respond and adapt to weightlessness. In the fish, the scientists examined bones, muscles, the nervous system, the spleen, and the vestibular organ. In the case of the algae, the interest of the researchers focused on proliferation, photosynthesis, and the light-guided swimming behavior of the mobile protozoa (Häder et al. 2018; Hilbig and Anken 2017) (Fig. 3.2).

The ModuLES Project (since 2011) Building on the experience gathered from the CEBAS and Aquacells/OmegaHab projects, the next generation of bioregenerative systems is now being developed in a modular approach—ModuLES (Modular Life Support and Energy System) (Wagner et al. 2016; Braun et al. 2018). Central element of the ModuLES system is a microalgae-based photobioreactor that has been technologically optimized for maximum efficiency and sustainable resource-saving production of oxygen and biomass. Both products can either be used directly or can be fed into other modules designed for various purposes, i.e., using the biomass for 3D-bioprinting of recyclable consumables like tools or cutlery, of tissue constructs for producing food supplements or other essential components like medication. Funded by the German Space Agency DLR, a completely closed and autonomous bioreactor system was developed by OHB and tested in parabolic flights. The bioreactor that recycles 96% of the water is complemented only with nutrients that have been consumed by the microalgae. Another unit efficiently separates algae and medium and facilitates highly efficient gas exchange. Autonomous operation is controlled by a sensor array measuring process parameters like pH, oxygen and biomass production, nitrate, phosphate, and sulfate. Within the scope of the ModuLES concept, it is foreseen to develop additional ModuLES subsystems and connect these to the microalgae photobioreactor in order to gradually evolve the modules into a complex bioregenerative life support system for research purposes and eventually for the supply of oxygen, biomass, and energy for exploratory long-term missions.

Despite the significant progress achieved in Europe in the development of bioregenerative life support systems over some 30 years, it is still a substantial way to go before such systems can reliably complement or substitute the currently used physicochemical systems.

3.3   A Side View to Astrobiology, Radiation Research, Microbial Challenges, and Protein Crystallization

Recent volumes in our SpringerBriefs Series on "Space Life Sciences" have described in detail the contributions of space research to these topics (Ruyters et al. 2017; de Vera 2020; Hellweg et al. 2020). Therefore, a brief side view should be sufficient summarizing the important findings.

3.3.1   Space Research in Astrobiology Helps to Unravel the Secrets on the Origin, Distribution, and Future of Life

How did life on Earth evolve? How did evolution take place? Does life exist anywhere else in the universe? Astrobiology investigates these fundamental human questions. Answering these complex questions requires an interdisciplinary approach that brings together aspects of cosmology, astronomy, planetary research, geology, chemistry, and biology.

Fig. 3.3 Various biological organisms and molecules were exposed to the conditions of space and a simulated Mars atmosphere mounted on the external Expose-R platform outside the European Columbus module of the ISS. (© ESA, Roskosmos)

In experiments led by DLR research institutes in cooperation with numerous international scientific teams, various organisms were exposed to the extreme environmental conditions of space on the outside of the ISS in order to investigate their ability to survive (for review see de Vera 2020). The results of the European Expose-E, Expose-R, and Expose-R2 missions—facilities attached to the outside of ISS—have been thoroughly summarized by Horneck and Zell (2012), Horneck et al. (2015), and by Cottin and Rettberg (2019) in three special collections of the journal *Astrobiology*. The overall aim is to answer the question of whether life originated on other celestial bodies and from there has spread to Earth (panspermia hypothesis). The results available so far confirm the enormous ability of microorganisms and spores as well as of lichens to survive—especially if they are protected by a layer of meteorite material or if they are associated with so-called biofilms (Venkateswaran et al. 2014a). Biofilms are structured communities of microorganisms on a surface, encapsulated in a self-developed matrix of extracellular polymeric substances. Dust particles contained therein form a protective shield, which additionally protects the organisms above all from the extraterrestrial ultraviolet radiation (Fig. 3.3).

Important discoveries, which have been made since the beginning of the new century and mainly achieved in ESA projects by European scientists, have been impressively highlighted in a recent publication (Cottin et al. 2015). Examples are the new understanding of planetary system formation including the specificity of the Earth among the diversity of planets, the origin of water on Earth and its unique properties among solvents for the emergence of life, and new concepts about how chemistry could evolve towards biological molecules and biological systems. Possibly, our planet may have been inhabited much earlier than previously imagined as indicated by life in the form of microfossils or life-indicative chemical imbalances found in the oldest rocks of our planet. At the same time, these achievements of the space projects form the basis for the optimization of the sterilization techniques of spacecrafts ("planetary protection"), which should land on the surface of Mars in the future to search for traces of life there. Two further publications provide recommendations for future experimentation in low-Earth orbit and beyond (Cottin et al. 2017) as well as on Earth (Martins et al. 2017). In addition, the current knowledge regarding the potential of life in the universe is summarized by Preston and Rothschild (2020).

3.3.2   Space Radiation: A Hazard for Astronauts in Exploration Missions

Space radiation—consisting of solar radiation, galactic cosmic, and extragalactic radiation—probably poses the greatest challenge for humans and materials when going on long-duration space travel to distant targets. The heavy, high-energy ions of galactic cosmic radiation are the greatest danger (for review see Hellweg et al. 2020).

Much attention has been paid to the radiation risk since the beginning of crewed spaceflight. During all space missions, measuring devices were flown to record the strength and composition of space radiation and to analyze their biological effectiveness. In fact, German research in space life sciences began with an experiment to measure radiation, namely with BIOSTACK on Apollo 16 in 1972. Since then, German expertise in this field has continued to develop, particularly due to the work of the DLR Institute of Aerospace Medicine and its many cooperation partners at German and international research institutions. Interestingly, the utilization of the ISS in the frame of the German life sciences program also started with an experiment on radiation measurement, namely with DOSMAP (Dosimetric Mapping) in spring 2001, a cooperation between DLR and NASA.

In the meantime, other devices that measure the radiation field inside and outside the ISS have been brought to the ISS. "Matroshka" is a phantom model of the human body used to investigate space radiation and its biological effectiveness on human organs and tissues, which proved to be very different in the various ISS modules (see, e.g., Berger et al. 2013). Fortunately, studies of chromosomal changes in the astronauts' lymphocytes showed that radiation exposure in the ISS is tolerable for the astronauts (Horstmann et al. 2005). For long-term exploratory missions to Mars and other distant targets however a more severe radiation load is forecasted necessitating further studies and the development of protection measures.

In 2008, ESA started the program IBER (Investigations into Biological Effects of Radiation) to study biological effects of space radiation and find solutions for effective radiation protection measures at GSI (Helmholtzzentrum für Schwerionenforschung in Darmstadt). IBER's objectives, scientific results, and perspectives, as well as further research plans in Europe for radiation health, have been summarized most recently (Walsh et al. 2019).

The knowledge gained from radiation research in space and on ground is not only helpful to assess the radiation risk for space travelers; it is also used in civil aviation, environmental protection, and health care. The investigation of the effects of individual heavy ions is particularly relevant for cancer therapy since the cellular responses to the different components of radiation are not yet fully understood. Also, successful developments of nutritional and pharmaceutical countermeasures—often going hand-in-hand between space radiation experts and clinicians engaged in cancer therapy—are of benefit likewise for astronauts and patients on Earth. Thus, the ongoing space radiation research is contributing to humankind's continuous search for remedy against cancer and aging (Hellweg et al. 2020).

3.3.3   Microbial Contamination: Challenges for Astronauts and Infrastructure

Microbial contamination—basically unavoidable in closed habitats—has been an issue also in spaceflight for decades. Despite strict quarantine procedures, studies in various spacecrafts indicate that a high diversity of bacteria, fungi, and actinomycetes are commonly carried onboard. They come from clothes, equipment, air currents during spacecraft handling and loading, food, and—not the least—from the astronauts themselves: the "microbial load" of human beings has been estimated to be in the order of 1.5 kg.

Microorganisms and especially biofilms, consisting of bacteria, yeasts, and fungi, can pose serious threats to space crews and the entire space infrastructure. On the one hand, they may impair the astronauts' health, while on the other hand hardware may be damaged because of their corrosive impact on a wide range of materials. Studies performed on MIR indicated that microbial damage to polymers and metals resulted in malfunctioning and even breakage of certain H/W units, e.g., of air conditioners and water recycling systems (Novikova 2004). In an early ISS study, more than 70 species of bacteria and fungi have been demonstrated to occur on surfaces (Novikova et al. 2006).

From this, it is not surprising that microbial monitoring has been a topic for space agencies and scientists over decades. The current status and some future perspectives have recently been nicely summarized (Yamaguchi et al. 2014). While the early Russian studies were unavoidably limited by their reliance on culturing to identify microbial species, more recent studies have implemented culture-independent approaches including small-scale 16S rDNA PCR surveys (e.g., Venkateswaran et al. 2014b; Ichijo et al. 2016; Lang et al. 2017a). The European "Extremophiles" study carefully assessed the microbial diversity distribution, functional capacity, and resistance profile on ISS. Great similarities were found between microbial communities on ISS and in ground-based isolation environments, which fluctuated but did not change significantly over time. And more importantly, although biofilm formation was detected and might become a critical issue with respect to material integrity and function, the results did not yet indicate a thread for human health (Mora et al. 2019; Voorhies et al. 2019).

Also on Earth, there is growing appreciation of the importance of microbial communities found in diverse environments from oceans to soil, to plants, and animals with expanding microbial ecology to human-constructed entities like buildings, cars, and trains—places where humans spend a large fraction of their time (Lang et al. 2017a). The long-standing experience of space research in this area may help to fertilize these efforts.

New methods for onboard microbial monitoring in space habitats aimed at detecting, quantifying, and identifying microorganisms of interest become increasingly important and are being developed by all ISS partners (Yamaguchi et al. 2014). The E-Nose system has been developed by the German industry to identify microbial contaminants by sniffing their volatile organic compounds with an array of metal oxide sensors on ISS (for details see Sect. 4.6.1).

The development of antimicrobial surfaces for facilities and infrastructure is another promising approach towards deep space exploration in the long term. In fact, researchers from Berlin together with the company Largentec have tested a new silver- and ruthenium-based antimicrobial coating aboard the International Space Station (ISS). AGXX® dramatically reduced the number of bacteria on contamination-prone surfaces (Sobisch et al. 2019; see also Sect. 4.6.2).

3.3.4   Protein Crystallization in Microgravity: Breakthroughs in Structure Analysis and Drug Design

The utilization of microgravity for improving protein crystallization and thereby structure determination started in the early 1980s. Theoretical consideration had led researchers to believe that, given the absence of sedimentation and gravity-driven convection in microgravity, crystals grown in space were likely to be larger and of greater purity. Although, as we know today, crystal growth is affected by a number of other factors as well, the assumption, on principle, had been correct. After the successful pioneering work by Littke and coworkers with experiments onboard of TEXUS sounding rocket flights in 1981 and during the first Spacelab mission in 1983 (Littke and John 1984) especially the German space life sciences program put much effort into this topic. Despite some technical and methodological drawbacks, early successes could be obtained as well. Especially after the development of the APCF (Advanced Protein Crystallization Facility) by German industry on contract by ESA and its utilization in Spacelab missions and on the International Space Station ISS the potential of microgravity for the improvement of crystallization and structure elucidation including possible applications became clearly visible. Since the book "Biotechnology in Space" in this SpringerBriefs series (Ruyters et al. 2017) devotes several chapters to this topic encompassing international achievements, only a few highlights achieved in the frame of the German program will be summarized here.

3.3.4.1   Crystallization of Ribosomes Awarded with the Nobel Prize

In October 2009, the Nobel Prize Committee awarded the Nobel Prize in Chemistry to two scientists working in the USA, Venkatraman Ramakrishnan and Thomas A. Steitz, and to Israel's Ada E. Yonath for their groundbreaking work on the structure and function of ribosomes. Yonath, who had worked in Germany between 1979 and 2004 as a guest professor at Berlin's Max-Planck-Institute for Molecular Genetics and later headed a Max Planck research group at the German Electron Synchroton (DESY) in Hamburg, was involved in a series of 12 space missions. The crystals grown during Shuttle/Spacelab missions D-2 (1993), IML-2 (1994), and USML-2 (1995) were larger, purer, and more evenly shaped thus guiding the way to further experiments on Earth, which then ultimately ended up in a successful elucidation of ribosome structures, and finally the Nobel Prize. Cooperation partners from DESY and from the FU Berlin continued—together with other colleagues—this research in microgravity, later also on ISS.

3.3.4.2   Structure and Function of the Photo System I

Two protein cofactor complexes, photo systems I and II, ensure that the biological energy conversion process during photosynthesis of green algae and plants runs extremely effectively. Therefore, structure, function, and dynamics of these complexes are important to be understood in detail. Based on crystals grown during the USML-2 shuttle mission in 1995 an improved structural model was developed by scientists from Berlin's Technical University at a resolution of 4 Å (or $10^{-4}\mu m$) making important functional parts of this large complex visible for the first time (Krauß et al. 1996). Further improvements with an even better resolution and fewer defects were obtained during the STS-95 mission in 1998 (Klukas et al. 1999).

3.3.4.3   The Crystallization of Archaea Surface Proteins

During shuttle flight STS-95 in October 1998, a team of scientists from the universities of Ulm and Mainz together with their Belgian colleagues succeeded for the first time in growing crystals of the S-layer glycoprotein of the archaebacteria *Methanothermus fervidus* (Evrard et al. 1999). So-called S-layers are probably the first cell wall structures that have evolved some 3 billion years ago showing an enormous resistance to heat, extreme pH values, and high salt concentrations. X-ray crystallography of these crystals and of those from later ISS experiments in 2002 led to a resolution of 3 and 1.9 Å, respectively, thereby providing a great step forward towards understanding the structure of the S-layer (Claus et al. 2002). These experimental results also enabled new applications such as the development of ultrafiltration membranes and molecular nanotechnology.

3.3.4.4   Mistletoe Lectin for Immune Stimulation and Cancer Treatment

The main component in mistletoe extracts, frequently used to strengthen the human immune system and to support cancer treatment, is mistletoe lectin-I. A closer investigation of its three-dimensional structure should help to provide clarification of its mode of action. Experiments on the ISS conducted between 2001 and 2006 enabled for the first time to explain the processes in the active centers (Krauspenhaar et al. 2002; Meyer et al. 2008). Built from two separate protein chains, called A and B, it is assumed today that subunit B is able to recognize certain sugar molecules on the membrane of the cell to be attacked, and thus helps subunit A to penetrate into that cell. This process then inhibits the cell's ribosome activity and ultimately leads to the death of, for instance, a cancer cell. Thus, space experiments have helped to understand the mechanisms of action thereby improving the pharmaceutical application of this protein.

3.3.4.5   Mirror-Image RNA Molecules

Ribonucleic acids (RNAs) provide the link between the DNA's genetic information and the proteins assembled by ribosomes as a result. Crystals of ribosomal 5S rRNA, an important component of the ribosomes at which protein synthesis takes place, were produced on shuttle flight STS-95 and onboard the ISS in 2001 and resulted in a detailed structure of domain B of the 5S rRNA. The model of 5S rRNA opens up new insights into the interaction between antibiotics and ribosomal RNAs thus permitting the development of more effective drugs (Vallaza et al. 2002).

Further experiments conducted in cooperation with the pharmaceutical company NOXXON AG on the ISS in 2002 led to the first successful crystallization of mirror-image RNA (Vallaza et al. 2004). The advantage of mirror-image nucleic acid compared to "natural" RNA molecules lies in their long life in the human blood. This makes them particularly suitable for an effective treatment of tumors or viral infections such as AIDS. In addition, thanks to their great stability, they can be chemically synthesized in large quantities and high levels of purity.

3.3.4.6   Outlook on Protein Crystallization in Space

While in Europe research is still attempting to disclose the mechanisms of the crystallization process in microgravity on ISS (Martirosyan et al. 2019), modeling the structure of key proteins grown under microgravity conditions is becoming a major focus of pharmaceutical companies in the US and Japan (for review see, e.g., McPherson and de Lucas 2015 and https://www.nasa.gov/sites/default/files/atoms/

Fig. 3.4 The German ESA astronauts Alexander Gerst working on a protein crystallization experiment on ISS. (© NASA)

files/benefits-for-humanity_third.pdf). By following the most recent recommendations from the US National Research Council to focus on compelling biological problems, where the production of diffraction-quality crystals on Earth is extremely challenging and preventing structural analysis, further positive results are to be expected from crystallization experiments in space with high economic value.

Recently, the Michael J. Fox Foundation launched an experiment to ISS with CRS-10 that involves growing high-quality crystals of a protein which plays a role in Parkinson's disease and could help to better understand its molecular structure and to design new efficient medication (Fig. 3.4).

3.4   ICARUS: International Cooperation for Animal Research Using Space—A New Field of Space Life Sciences Research

A completely new field of space life sciences research has been opened up by the ICARUS initiative. ICARUS collects data on the global movements and the migration of animals in real-time. However, the telemetry data received by an ISS antenna from tiny transmitters attached to all kinds of animals like insects, birds, cattle, rhino, and elephants will also become a valuable source of information on the state of our planet Earth. Since animal senses are often superior to human senses, the information

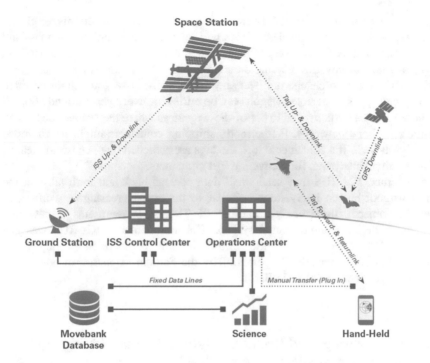

Fig. 3.5 The migration and local movements of animals as well as environmental data are transmitted by small tags via the ISS antenna to a ground station and the Movebank Database. (© DLR)

could be particularly useful for predicting catastrophes, the spreading of infectious diseases but also for monitoring ecological and climate changes.

Under the leadership of the Max-Planck-Institute of Animal Behavior in Radolfzell, Germany, scientists from Russia, the US, and various European countries are working together to analyze the telemetry data of the animal transceivers called tags (https://www.icarus.mpg.de/en). The tags weighing only 5 g include battery, GPS module, accelerometer, and magnetometer as well as temperature, pressure, and humidity sensors. The data is decoded by a receiving antenna on the ISS and transmitted to the ground station in Moscow. From there, the data is transferred to Movebank, a global database for animal movements. By using the ISS as an orbital relay station, the ICARUS system is not dependent on terrestrial communication networks which are currently used in the most common equipment for animal tracking and are often unreliable for sparsely populated areas.

In the future, the new technology can be used not only to answer important biological questions—such as where fruit bats fly to in Africa and where they encounter Ebola viruses or where sea turtles spend their youth and where most of them die (Fig. 3.5).

From the combined information of thousands of animals, a synopsis of life on Earth is created, the so-called Internet of animals. It could also be used to transmit other technical or environmental data such as tree growth, ocean currents, or glacier melt. Measuring sensors attached to trees will precisely determine the biomass and thus the carbon dioxide turnover of forests. So, scientists hope that it will assist them in making new discoveries relating to the protection of our natural resources.

ICARUS can also change our everyday lives. Railway wagons might be fitted with special tags so that stolen wagons can be retrieved. Every year, around 100,000 of these vehicles are conveyed to out-of-the-way areas where they cannot be located by mobile phone networks. Furthermore, shipping containers might be equipped with tags as well. If a container is lost, the tags are switched on—a cost-efficient as well as redundant system for recovering lost containers.

Altogether, ICARUS may send small data packets back and forth all over the world without human intervention—similar to the ever-increasing networking of everyday objects, the so-called Internet of Things. It represents a new and completely digital communication system. The importance of ICARUS was just recently also recognized by the award in the competition "Bridges for German-Russian University Cooperation" by the Foreign Ministries of both countries on September 15, 2020, in Berlin.

3.5   Key Findings and Breakthroughs in Human Physiology and Space Medicine

During evolution humankind has adapted optimally to gravity; in fact, we often do not even notice its influence. Upright walking seems completely natural to us; but, basically, it is a continuous battle with gravity, which becomes obvious when thinking of babies learning their first steps or of elderly people struggling with getting up or walking steadily. During upright walking, which involves two-thirds of the total skeletal musculature—especially back, hip, and leg muscles—about 90% of the energy used is necessary to keep the body position upright and to lift the legs against gravity, while only 10% of the total energy is necessary for the movement of the legs, i.e., for the locomotion as such. In our younger age, muscles and bones are well suited to cope with gravity. The same is true for the cardiovascular, the vestibular as well as the immune systems. During aging however problems arise: Roughly at the age of 40 already, musculature starts to degrade, and walking becomes increasingly bowed. Muscle and bone mass decrease further at old age, getting up from a chair often needs to be supported by our arms, and later gravity ties us literally to bed. Also, all other physiological systems of the human body degenerate with a more or less recognizable effect.

Especially in the mobile societies of the (Western) industrialized world, aging is one of the great challenges for the upcoming decades. In Germany, people above 65 years will represent roughly 29% of the population by 2030 (today 20%), 2060 about 34% (www.demografische-forschung.org). In its visionary position paper on the so-called "High-Tech-Strategy of the German government" the German

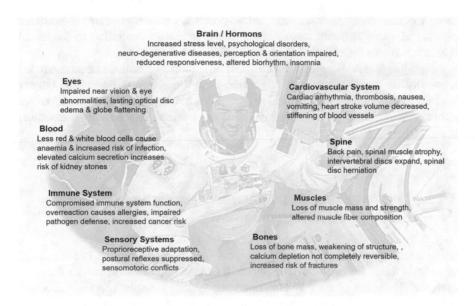

Brain / Hormons
Increased stress level, psychological disorders,
neuro-degenerative diseases, perception & orientation impaired,
reduced responsiveness, altered biorhythm, insomnia

Eyes
Impaired near vision & eye
abnormalities, lasting optical disc
edema & globe flattening

Cardiovascular System
Cardiac arrhythmia, thrombosis, nausea,
vomitting, heart stroke volume decreased,
stiffening of blood vessels

Blood
Less red & white blood cells cause
anaemia & increased risk of infection,
elevated calcium secretion increases
risk of kidney stones

Spine
Back pain, spinal muscle atrophy,
intervertebral discs expand, spinal
disc herniation

Immune System
Compromised immune system function,
overreaction causes allergies, impaired
pathogen defense, increased cancer risk

Muscles
Loss of muscle mass and strength,
altered muscle fiber composition

Sensory Systems
Proprioceptive adaptation,
postural reflexes suppressed,
sensomotoric conflicts

Bones
Loss of bone mass, weakening of structure, ,
calcium depletion not completely reversible,
increased risk of fractures

Fig. 3.6 Major effects of spaceflight on the human body. (© M. Braun, photo NASA)

government provided special attention to the topic of health protection and health care. From this, it is not surprising that the document identified the theme of "Health and Nutrition" as one of the five global challenges with a special focus on aging research (www.BMBF 2010).

Space medicine here is asked to contribute its findings to meet this challenge since astronauts must battle with very similar health problems in space as aging people on Earth. In space we see—due to the lack of gravity and minimal levels of movement and resistance—a drop in the performance of the cardiovascular system, disturbances in the vestibular and immune systems, muscular atrophy, and bone degradation (for a recent summary see Norsk 2020). In contrast to the aging process, the changes experienced by astronauts occur within days or weeks thus offering a time-lapse view of this process on Earth. Luckily, most of the changes are, unlike the aging process, reversible, and space medicine can take great benefit from investigating the readjustment process of the human body to terrestrial gravity conditions (Fig. 3.6).

While at the beginning of human spaceflight, space medicine was established and focused on maintaining the health and performance of astronauts, it very soon became clear that people on Earth highly benefit from the results from space medicine and life sciences research in general. A recent review summarizes important results from space physiology in context with aging and terrestrial medicine and calls for a synthesis of knowledge achieved by studying microgravity-induced deconditioning in space and that during bed rest confinement. The more comprehensive and integrative approach usually followed in space medicine research incorporates aspects such as nutrition, muscle and bone strength and function, cardiovascular deconditioning, and cardio-postural interactions can provide new insights of value for both space medicine and geriatrics according to the motto "Geriatrics meets Spaceflight" (Goswami 2017; Vernikos 2020).

In adapting these thoughts, we will focus now on major achievements generated especially in the European and German space life sciences programs. Further selected discoveries by international experts from human research in space that are relevant to human health on Earth have been summarized very recently (Shelhamer et al. 2020).

3.5.1   Cardiovascular Regulation: Astronauts and Patients Benefit from New Approaches

Immediately after entering microgravity conditions in space, about 2 l of blood and other body fluids are shifted to the upper half of the body of astronauts. At the same time, the regulation of the autonomic nervous system changes, and the cardiovascular system gradually adapts to the lack of gravity. When an astronaut returns to Earth, these systems are abruptly confronted with gravity again. Consequently, circulatory regulation disorders are frequent in the first few days—similar to the orthostatic symptoms encountered by elderly people when they want to get out of bed too quickly in the morning. Especially at the beginning of human spaceflight large efforts were therefore undertaken to monitor and stabilize the functioning of the cardiovascular system. Not rarely, these efforts led to surprising results, some of them even in contradiction to textbook knowledge.

3.5.1.1   Contradicting Textbook Theory: Central Venous Pressure Decreases in Space

Experiments on the first Spacelab mission in 1983 and on the German D-1 mission in 1985 focusing on blood pressure measurements yielded surprising results. As mentioned, microgravity induces a rapid fluid shift in astronauts from the lower to the upper part of the body. In theory, this hypervolemia was expected to increase venous pressure simply because the vessels are fuller. Thus, it came as a big surprise that indirect measurements of central venous pressure (CVP)—the pressure in the venae cavae—showed a significant decrease during the FSLP (1983) and D-1 (1985) Spacelab missions (Kirsch et al. 1984, 1986). Only after direct measurements of CVP by a venous catheter on D-2 in 1993 and following shuttle flights (Foldager et al. 1996), the scientific community started to believe in these data and accepted the unexpected result—and, by the way, it took several years before these data could be explained after further experiments on parabolic flights that demonstrated an even greater decrease in intrathoracic pressure than in CVP (Videbaek and Norsk 1997).

These findings from space missions and parabolic flights, complemented by data from bed rest studies, are also relevant in context with the treatment of hypotension and in decreasing intracranial pressure by using an inspiratory threshold device (Convertino et al. 2011). These medical advances would not have been possible without the early measurements from the Spacelab missions.

3.5.1.2   Transient Rise in Intraocular Pressure: No Challenge for Astronauts

As mentioned above, cardiovascular problems were assumed to be a serious challenge for astronauts in the early days of spaceflight. In this context, there was concern in operational medicine in the 80s and early 90s that fluid shift may also cause an increase in the intraocular pressure of astronauts; in fact, elevated intraocular pressure is known to cause glaucoma in humans. As preliminary data from the D-1 Spacelab mission seemed to support this concern, a so-called self-tonometer was developed by Prof. Draeger from the Eye Clinic of the University of Hamburg in order to measure intraocular pressure noninvasively and as quick as possible after the astronauts enter into microgravity (Groenhoff et al. 1992; see also Sect. 4.2.1). Data from the MIR'92 and D-2 space missions in 1992 and 1993, respectively, taken a few minutes after reaching microgravity indeed demonstrated a rapid and nearly twofold increase of intraocular pressure; fortunately, however it decreased again in the course of some hours and returned to the normal preflight level after 2 or 3 days (Draeger et al. 1993, 1994). From these results, severe consequences for the eye health of astronauts could be ruled out.

Interestingly however the measurement of intraocular pressure has got renewed attention lately in context with SANS (spaceflight associated neuro-ocular syndrome). Under this term, unusual physiologic and pathologic neuro-ophthalmic findings in astronauts such as visual acuity impairments are summarized. Various hypotheses have been proposed for explanation such as increased intracranial pressure, but few data exist up to now (Lee et al. 2020a). To shed more light on this important topic, many studies on SANS are presently being performed or are in preparation on ISS and in terrestrial analog environments (e.g., bed rest studies).

3.5.1.3   Significant Adaptation Capability of the Cardiovascular System

Over the years, cardiovascular issues and events were always carefully monitored by the crew doctors and seemed to be largely under control also due to proper exercise regimes. A continuous monitoring is, nevertheless, still required and performed also on the ISS. In this context, scientists from the Hanover Medical School—in the meantime working at the DLR Institute of Aerospace Medicine in Cologne—hoped to learn more about the causes of circulatory changes in astronauts intending to use the results also for new diagnostic and treatment concepts on Earth. In collaboration with Russian scientists from IBMP, the ISS experiments PULS and

PNEUMOCARD were performed using portable, reliable devices for the noninvasive monitoring of the cardiovascular system specially developed for space application (Baevsky et al. 2009; Tank et al. 2011; see also Sect. 4.2.4). Between 2002 and 2012, measurements were conducted on more than 30 cosmonauts in cooperation with the Institute of Biomedical Problems in Moscow. In the experiment "Pulse" (2002–2007), ECG, pneumotachogram, and peripheral pulse were registered with the help of a finger photoplethysmography sensor, for "Pneumocard" (2007–2012), seismic cardiography and impedance cardiography were added. The results showed that cardiac function, blood pressure, and heart and respiratory rates adapt to space conditions amazingly quickly and well. After the cosmonauts' return to Earth, their cardiovascular systems stabilized within 3 days—evidence of the effectiveness of the exercise programs on the ISS as well as for the remarkable adaptation capability of the healthy human cardiovascular system (Baevsky et al. 2014). This is in perfect agreement with a recent NASA statement that crew members who maintain an appropriate dietary intake coupled with adequate, programmed exercise, sleep, work, and recreation periods can maintain good cardiovascular health and well-being.

On the other hand, comparative studies of patients with orthostatic problems on Earth have surprisingly shown that some patients—unlike astronauts—often do not benefit from physical activity. In these cases, genetic and environmental factors may play a part. This might possibly lead to a new approach to treating these patients.

3.5.1.4   Two-Component Model of Blood Pressure Regulation Supported by Microgravity Research

Cardiovascular diseases are still one of the main causes of death in the Western world with high blood pressure playing a dominant role. Two further aspects of cardiovascular regulation have recently come to the attention of the researchers. In addition to the peripheral arterial blood pressure, a well-established factor in human physiological research as well as in the treatment by medical drugs for decades, the central aortic blood pressure has now got increasing attention. This is due to the fact that evidence has grown that central aortic pressure may be a more powerful predictor for cardiovascular events. However, conclusive data on its regulation were rare with data in microgravity completely lacking. For the first time, the recent development of cuff-based oscillometric instead of tonometric pulse wave analysis technologies allowed an assessment of central aortic pressure within a few seconds also in microgravity.

In a precursor experiment on parabolic flights, a differential effect of changes in gravity conditions on central versus peripheral blood pressure could be demonstrated supporting the two-component model of blood pressure regulation. While the peripheral blood pressure remained largely constant, the central pressure increased significantly (Seibert et al. 2018). If this would be true also for long-term microgravity, the cardiovascular risk profile in astronauts, but also in patients on Earth may need to be reinvestigated.

3.5.1.5 Microcirculation Monitoring by Noninvasive Vital Microscopy: A New Approach to Detect Circulatory Problems

Another new aspect that has come into focus in human physiological and clinical research is the investigation of the role of microcirculation, the blood flow in the smallest human vessels. It has great significance for the human organism as an important blood reservoir, and affects blood pressure, promotes heat exchange, and transports oxygen and vital nutrients to cells. Basically, the whole cardiovascular system relies on efficient microcirculation. Since microgravity affects the cardiovascular system, monitoring of microcirculation could give important insights into the health and well-being of astronauts (Bimpong-Buta et al. 2018).

One of the modern ways to measure microcirculation is the so-called intravital microscopy using the Microscan microscope. The device—a special manual microscope the size of a smartphone that takes measurements of blood circulation in capillaries at the base of the tongue in real-time—has been developed a few years ago by a small spin-off company, originating from the Dept. Clinical Physiology, Academic Medical School of the University of Amsterdam (www.microvision. medical).

Experiments analyzing the effects of varying gravity conditions were performed in parabolic flights in 2017/2018. Dependent on the positioning of the test subjects changes in macro but not in microcirculation were demonstrated supporting the view that the two processes do not necessarily correlate (Bimpong-Buta et al. 2020). If this holds also in long-term microgravity, consequences for the positioning of astronauts during their operational work would need to be taken. The findings from the parabolic flights are thus calling for the monitoring of microcirculation and by this for the development of new diagnostic alternatives for astronauts on ISS and upcoming exploration missions. At the same time, this new approach could help to identify people on Earth with an increased risk of circulatory problems with a simple noninvasive method.

3.5.1.6 "Space Fever": Critically Elevated Body Core Temperature in Exercising Astronauts

As already mentioned, fluids such as blood and lymph quickly move from the lower to the upper part of the human body after the onset of weightlessness. Corresponding changes in heat regulation occur at the same time: astronauts often complain about cold feet and fingers. On the other hand, hard work in a spacesuit may very quickly lead to a rise in temperature of the internal organs—the so-called body core temperature—to the dangerous level of 39 °C and above, the cause probably being the impaired convective heat transfer and efficiency of evaporation in microgravity. Monitoring the body core temperature is thus a must.

Exactly for this purpose, the noninvasive temperature sensor described in more detail in Sect. 4.2.3 had been developed and was used for the "Thermo" and

"Circadian Rhythm" experiments on ISS from 2009 until 2019 (Gunga et al. 2016). The double sensor was initially tested on parabolic flights. In addition, scientists used a thermal imaging camera to register temperature distribution as well as other modern methods to detect rapid movements of liquid under changing gravity conditions. Not surprisingly, more than half a liter of blood rises from the legs towards the head immediately after the onset of weightlessness. Measured noninvasively in parallel for the first time, heat radiation—around 30% of which is emitted by the head even under normal conditions—increases remarkably in weightlessness.

On the ISS, measurements that ultimately involved a total of 12 astronauts were performed on the ISS between 2009 and 2012. Results of the "Thermo" experiment surprisingly show that in microgravity the basal temperature at rest is higher by about 0.5–1 °C than under terrestrial conditions. What is even more important is that body core temperatures are relatively high during physical training in space and will stay that way for a much longer period, even during rest phases after exercise. In some cases, a critical temperature of above 40 °C was reached. Only several weeks after return to Earth the thermoregulatory system of the astronauts achieved the same performance level as before flight. In the future, these findings will help mission planners improve their estimates of the astronauts' workloads and of the rest periods required and protect them from overheating (Stahn et al. 2017). In view of the expected climate changes on Earth, these findings also raise concerns about the work efficiency and evolutionary ability of human beings to adapt to a higher temperature (Gunga 2020).

The follow-up experiment on the ISS called "circadian rhythm" performed between 2012 and 2019 aimed to investigate how long-term missions affect the biological clock by measuring the body core temperature and the level of melatonin, a hormone that is intricately connected to the circadian rhythm. This biological clock that gives us a sense of what time of day it is and makes us get tired at night is normally synchronized with the Earth's 24-h rotation period. Measurements in space performed in the 1990s already pointed to a disturbance of sleep and the circadian rhythm (Crandall et al. 1994; Gundel et al. 1997; Dijk et al. 2001). Preliminary results from the ISS experiment indicate an acceleration as well as an increase of the temperature maxima. Since any change in the circadian rhythm has a negative influence on people's quality of sleep, their alertness, and their intellectual performance and their mood, the results hopefully will tell us how people can get to rest most effectively. Thus, not only astronauts stand to benefit from these results but also shift workers and other individuals working irregular hours. Their circadian rhythms are often disrupted, and this study could improve our understanding of how the body's core temperature adapts to a changing internal clock.

The project "Spacetex" added results to the scientific experiments on temperature regulation. With the lack of convection in space, the usual evaporation or dripping of sweat during sports cannot take place, leading to an uncomfortable feeling as well as the risk to overheat the body during exercise since the cooling effect of evaporation is missing (Stahn et al. 2017). Various attempts have been made to develop special new fabrics that allow to control the thermal and sweat management in order to maintain the astronauts' cooling mechanisms and reduce microbiological contamination in the ISS. In the project "Spacetex-2," the Hohenstein Institute, Schoeller Textil AG, and the Charité Berlin have partnered to develop and test special garments for the ISS. It is based on the results of Spacetex-1 performed during the "Blue Dot" mission of the German astronaut Alexander Gerst in 2014. For his mission "Horizons" in 2018 functional Spacetex T-shirts were specially designed and fabricated to the astronaut body. The garments consist of a top and pants consisting of PES, PET, and PBT or of cotton, respectively (www.hohenstein.de). After utilization on ISS, the T-shirts were sealed and brought back to Earth for analysis of contamination and odor generation. It is expected that this special design will contribute to the astronauts' comfort and well-being—especially important for long-duration exploratory missions.

3.5.1.7   Textbook Knowledge of Pulmonary Ventilation and Perfusion Questioned by Electrical Impedance Tomography (EIT)

In addition to the cardiovascular system, also ventilation and lung function, in general, have been thought to be affected by microgravity. Based on theoretical considerations, scientists had assumed for a long time that pulmonary ventilation and perfusion had to be completely homogenous in weightlessness. However, early experiments in space—during the German D-2 shuttle mission in 1993, for example—had shown that this is not quite correct (Verbanck et al. 1997). Consequently, research teams around the world started to address this problem over several years (for review see Prisk 2014; Gunga et al. 2016).

On parabolic airplane flights in 1999 and 2000, an improved version of an old method—Electrical Impedance Tomography (EIT; see Sect. 4.2.2)—was applied (Frerichs et al. 2001). For the first time, researchers were able to identify gravity-related regional changes in the function of the lungs as well as shifts in the volume of blood and liquid in the chest. Indeed, some residual inhomogeneity remained that is today explained as being a direct consequence of the coupling of ventilation and perfusion (see Shelhamer et al. 2020 for more details). Moreover, experiments on parabolic flights as well as accompanying studies on the ground revealed that the pattern of air and blood distribution in the lungs that is described in textbooks does not apply to all people (Frerichs et al. 2001). Because the influence of gravity on this distribution plays an important part particularly in diseased lungs, these discoveries acquire clinical significance in therapy management during artificial respiration as well as in diagnosing neonatal respiratory failure and the so-called adult respiratory distress syndrome (ARDS; Frerichs et al. 2010).

3.5.2   Challenges for the Vestibular System, Balance Control and Motor Coordination in Microgravity

3.5.2.1   Space Motion Sickness and the Role of Otolith Asymmetry

The absence of gravity creates substantial perturbations in the vestibular system and more specifically in the functioning of the otolith organs and the semicircular canals in the inner ear. Normally, this system is providing information on the position and the movement of the head in the three-dimensional world, while the visual system, on the other hand, supplies general information about the relative position and movement of our body within the visual space. Thus, for instance, we are able to see whether or not a picture is hanging straight even if we have no information about our orientation relative to the force of gravity. At the same time, we have no problems standing upright in a completely dark room because the equilibrium system permanently identifies the alignment of the head in relation to gravity. In addition, touch sensors contribute to our overall spatial orientation on Earth. For this so-called proprioception, receptors are located in the skin, muscles, and joints to build the internal sense of our body.

Astronauts often suffer orientation problems in space—so-called space motion sickness (SMS)—due to a mismatch of information resulting from the vestibular system in the inner ear, the eyes, and the touch senses. Despite a lifetime of experience with normal Earth gravity, astronauts adjust—luckily enough—to this missing otolith signal within several days in space (Reschke et al. 1994). Aspects of this neuroplasticity have later helped to create training programs to improve motor function in clinical populations (Edgerton et al. 2000).

To elucidate the causes and mechanisms involved in space motion sickness, scientists from Charité Berlin aimed to study the function of the vestibular system in microgravity over many years via measurements of compensatory eye movements. In this context, several generations of eye movement tracking devices had been developed, finally the 3D-Eye Tracking System (3D-ETD) for the ISS (see Sect. 4.3)—creating surprising observations. So, in contrast to the theory of Nobel laureate Barany the caloric nystagmus can also be elicited in microgravity. Nystagmus is the repetitive reflexive motion made by the eyes to maintain gaze stability when the head is rotated. As a test of vestibular function, it can also be stimulated by irrigating the ear with hot or cold water—caloric stimulation, which sets up convection within the fluid of the semicircular canals thus mimicking an extended head rotation. Since in space there is no convection, the existence of caloric nystagmus measured in Spacelab experiments and on the MIR space station was surprising (Scherer et al. 1986; Clarke 2017). Later studies determined a direct thermal effect of caloric stimulation in the vestibular system, which slightly expands the canal membrane and induces fluid motion within the canal—results leading to a better understanding of this very commonly used diagnostic tool.

Experiments were continued on ISS with the newly developed 3D-ETD. From these experiments, complemented by research during parabolic airplane flights, it was found that the basics of signal processing in the brain achieved by training and experience is modified as gravity conditions change. Apparently, gravity influences not only the system of equilibrium but also the control of eye movements, in general, and thus possibly the visual process itself. In addition, the results provide evidence of a dominant labyrinth making a reconsideration of the otolith asymmetry hypothesis necessary: while on landing day after spaceflight responses from one labyrinth were equivalent to preflight values, those from the other showed a considerable discrepancy. The otolith asymmetry hypothesis, based on mass differences between the otoconia of the right and the left ear, cannot accommodate such a difference in adaptation. The results thus support the idea of a unilateral dominance in the utriculo-ocular neural circuitry as the primary factor rather than a morphological asymmetry as contributing to inflight and postflight disorientation (Clarke 2008; Clarke et al. 2013; Clarke and Schönfeld 2015), making a reconsideration of the otolith asymmetry hypothesis necessary. In general, findings here demonstrate that the gravity vector represents a common reference for vestibular and oculomotor responses. They also support the idea that the gravity vector provides a central reference for the entire sensorimotor complex—as reflected in perception and motor control in addition to the basic vestibulo-ocular, vestibulo-spinal, and vestibulo-autonomic mechanisms (Clarke 2017).

It should also be mentioned that in parallel to and based on the spaceflight experiments several clinical studies were performed directed at the improvement of clinical diagnosis procedures for suspected disorders of the vestibular function. Clarke (2017) provides a valuable summary of these achievements in clinical testing.

3.5.2.2 Motion Sickness, Stress, and the Endocannabinoid System (ECS)

Motion sickness experienced during road, air, sea, and space travel can be extremely debilitating, and yet the present understanding of its neurobiological mechanisms is incomplete. The traditional sensory conflict hypothesis described above suggests that motion sickness results from a conflict between actual and anticipated signals from sensory organs involved in spatial orientation. On the other hand, it is known that motion sickness with nausea and vomiting is linked to a pronounced activation of stress response systems also suggesting a possible involvement of the ECS in the gut-brain interaction. The role of ECS in humans in context with motion sickness had never been investigated, however. Experiments on parabolic flights clearly indicated that stress and motion sickness are associated with impaired endocannabinoid activity. Enhancing ECS signaling may thus represent an interesting alternative therapeutic strategy for motion sickness in individuals including astronauts who do not respond to currently available treatments (Choukèr et al. 2010).

3.5.2.3   Human Locomotion and Balance Training Under Varying Gravity Levels: Therapy Options for Elderly People and Patients with Motor Impairments

Prolonged stays of humans in space also lead to challenges in posture and locomotion provoked by structural and functional degenerations within the neuromuscular system, which is of major concern to space agencies regarding long-term interplanetary spaceflights in the future. A recent review provides an excellent summary of the mechanisms and consequences of changes in posture and locomotion induced by modified gravity conditions (Ritzmann et al. 2017). The authors demonstrate that—with an emphasis on motor learning and neural plasticity for cyclic locomotor movement as well as the control of posture—space research helps to gather knowledge about the neuromechanic mechanisms underlying movement control. Specifically, findings about the gravity-dependency of reflexes, sensory integration, and the interplay of muscle coordination and biomechanics provide valuable information relevant for a basic understanding of postural control and locomotion of astronauts as wells as in elderly people or patients in the clinic on Earth. An impressive example should further illustrate the importance of this space-related research for rehabilitation—the proof of the effectiveness of balance training with partially unloaded body weight for the improvement of postural control in elderly people and patients with motor impairment.

It is well-known that balance training improves postural skills and subsequently reduces the risk of lower limb injuries as well as of sustaining a fall. Consequently, balance training is used in therapy to counteract circumstances where postural and motor control are compromised. However, people suffering from motor impairments or reduced mobility due to surgeries, neurological diseases, or elderly people may not be able to bear their full bodyweight but may be possibly capable of performing training with partially unloaded bodyweight. This idea was tested by a research team at Freiburg University.

Partial unloading was achieved on Earth either by means of a harness attached to a height-adjustable system or during parabolic flights with an airplane flying special maneuvers providing not only microgravity phases but also Moon (0.16 g) and Mars gravity (0.38 g). The results clearly demonstrate that balance training under reduced load such as Mars gravity affects postural and motor control equivalent to training under 1 g conditions. Thus, balance training under partial unloading is an appropriate alternative for patients unable to conduct training under full body load as well as an appropriate countermeasure for preserving postural skills and for reducing negative effects caused by prolonged stays in microgravity (Freyler et al. 2014; Ritzmann et al. 2015, 2016). For this success, which now benefits hospital patients on Earth, the Freiburg research group was honored with the renowned Clinical Award of the American College of Sports Medicine.

Recently, investigations have utilized the varying gravity conditions during parabolic airplane flights to analyze the neuromuscular control of postural responses during bouncing and stumbling. Since the first steps of astronauts on the Moon, human locomotion, stance, stumbling, and bouncing have come into focus of investigations preparing future exploration missions. Proper coping with lunar or Martian gravity after a transit in microgravity is, of course, mandatory to perform mission-critical tasks as well as to reduce risks of injury on planetary surfaces. Luckily for astronauts and humans on Earth, first results of the parabolic flight experiments demonstrate a remarkable adaptability to compensate for the sudden deterioration of postural balance (Ritzmann et al. 2016, 2019; Holubarsch et al. 2019).

3.5.3   Challenges for Muscles, Bones, and Cartilage of Astronauts in Space

The musculoskeletal and cartilage system is of utmost importance for human locomotion and posture. Maintaining its integrity is thus essential for humans on Earth as well as for astronaut health during and after the mission. In fact, loss of bone and skeletal muscle in response to microgravity exposure has been a medical and physiological concern since the early Gemini, Soyuz, and Skylab missions. Today we know that the space-induced changes include extensive bone loss, loss of muscle mass and strengths, increased risk of kidney stones, vertebral disc alterations, lower back pain as well as changes to the elasticity of the tendons. Data from MIR, Shuttle/MIR missions, and the ISS reveal that especially weight-bearing bones have a continuous loss of areal bone density of about 1–2% per month. Thus, in roughly 1 month of spaceflight, astronauts experienced a loss of bone mineral density typically observed over a year in a postmenopausal woman (Bloomfield 2020). Since a few years, also the changes of cartilage have come to the attention of scientists—an aspect largely neglected in the past although cartilage recovery after its degradation is probably impossible in humans.

Since spaceflight provides a highly controlled environment with healthy individuals and a high degree of monitoring, it is an ideal setting in which to analyze bone loss and to test the effects of countermeasures in a kind of time-lapse mode (Shelhamer et al. 2020). Although the understanding of the mechanisms of muscle and bone loss is still incomplete despite decades of research in space and on Earth (Rennie et al. 2010), a wealth of information has been obtained in space experiments and accompanying bed rest studies. The results are of fundamental importance not only for sustaining human presence in space and enabling exploratory missions but also for the understanding of disuse-atrophy in bedridden patients and its contribution to sarcopenia in older individuals as well as for that of osteoporosis (Narici and de Boer 2011; Lang et al. 2017b).

Countermeasures against muscle and bone degradation have been developed and tested extensively especially in bed rest studies over the last decades, organized mainly by NASA and in Europe, here primarily by ESA, DLR, and CNES. Certainly, the development of effective countermeasures and therapies is of high priority in space agencies around the globe, also because recovery after flight is not as rapid and possibly incomplete in astronauts (for review see Narici and de Boer 2011). In fact, it is mandatory to continue these efforts especially in preparation of long-term exploratory missions, the more so, since countermeasures applied in space on the ISS are not fully satisfactory up to now. For the time being, new approaches such as whole-body vibration and exercise via a sledge jump system seem to be most promising. These examples will be described in Chap. 4 (see also Konda et al. 2019).

3.5.3.1   Movement Behavior and Muscle Atrophy Rather than Spine Elongation Causes Back Pain in Astronauts

In the past, more than half of the astronauts have complained about back pain during their spaceflights. However, its real cause remained unclear; spine elongation due to the lack of gravity was discussed as one possible factor. To address this issue a team from the German Sports University in Cologne started a project in the 1990s to measure not only changes in spine length but also the movement behavior of astronauts in space. A respective device for ultrasonic long-term monitoring of spine movements was developed in cooperation between the scientists from the German Sports University and an engineering team from the University of Jena (see Sect. 4.4.1).

The device was validated during various bed rest studies jointly performed by the French space agency CNES and DLR in the mid-90s. Based on the promising results, the respective flight H/W was developed and used on the Russian space station MIR during the Russian/German MIR'97 and—again in cooperation with CNES—during two French MIR missions in 1998 and 1999. The data showed that especially the specific movement behavior of astronauts in microgravity rather than the spine elongation itself is the main cause for the reported back pain (Baum et al. 1997). In bed rest studies, a specific movement regime of the test subjects indeed resulted in relief from back pain interesting for astronauts in long-duration missions, but also for humans on Earth (Baum and Essfeld 1999). Its application potential ranges from rehabilitation and physiotherapy, occupational health including work-place planning to recommendations of sitting positions of medical doctors and pilots, truck drivers to name only a few (see Sect. 4.4.1).

The results are also interesting in context with disc herniations. An increased risk of disc herniations in the lumbar as well as in the cervical spine of astronauts has recently been demonstrated to occur (Johnston et al. 2010). A detailed analysis of the astronaut data revealed that this risk shows a daily rhythm with being highest in the morning after waking from an overnight rest. On Earth, direct studies on diurnal variations are missing, but data on backpain lead to corresponding results. So, the astronaut data on herniations tell patients on Earth to carefully control spinal flexion and omit heavy lifting activities in the early morning to reduce the risk for hernia-tions (Belavy et al. 2016). In a more recent study, MRI data and dynamic fluoros-copy of the lumbar spine of astronauts after 6 months of microgravity exposure on ISS were presented suggesting that backpain and an elevated risk of postflight disc herniation result from muscle atrophy rather than from intervertebral disc swelling (Bailey et al. 2018; Elsevier Outstanding Paper Award Winner).

3.5.3.2   Research in Space Contributes to Elucidating Mechanisms of Bone Degradation in Space as Well as in Osteoporosis on Earth

According to Wolff's law, bones adapt according to the loads placed upon them. Decreased stress on bones due to microgravity thus leads to more resorption and less preservation. The most pronounced deconditioning of bones is localized in the lower limbs of astronauts. In particular, bone resorption in the trabecular compartment is higher than in the cortical compartment. Interestingly, this pattern of bone loss is reflected in osteoporotic patients on Earth. Following the return of astronauts to Earth, recovery of lost bone occurs very slowly; within the trabecular compartment, the loss may be permanent. This information is important also for terrestrial clinicians attempting to rehabilitate patients after significant periods of disuse, e.g., after amputation, lower-leg fractures, ligament tears, and other clinically required but temporary disuse conditions leading to bone loss with delayed or incomplete recovery (Shelhamer et al. 2020 and references therein).

Current theories suggest that normal remodeling rates are disturbed with bone resorption by osteoclast activity occurring at a faster rate than ossification by osteoblast activity. Mechanistically, systemic markers of bone resorption and urinary calcium excretion are consistently elevated, while bone formation markers are mostly unchanged leading to the result that remodeling is elevated and the dynamics uncoupled. The same alterations in remodeling are well documented in most bone diseases. Development of effective countermeasures is thus not only beneficial for astronauts but also for people with disuse-related bone changes, e.g., after amputation, lower-leg fractures, or other (temporary) mobility problems of the aging society (see Sect. 4.7). The finding that an antiresorptive agent, alendronate, was shown to be effective in reducing bone loss, remodeling rate, and calcium loss in recent bed rest studies indicate that early pharmacologic intervention at the beginning of a disuse period could probably preserve bone integrity also on the clinical setting.

In addition, nutrition is critical for health in astronauts and humans on Earth. Given the fact that bone loss during spaceflight is approximately ten times faster than on Earth and that spaceflight provides a highly controlled environment with healthy individuals and a high degree of monitoring it appears to be an ideal experimental setting for analyzing the mechanisms of bone loss as well as the efficacy of countermeasures including nutrient supplementation (Sutton 2020). Thus, the space environment, indeed, holds the potential to provide detailed insights into the therapeutic or detrimental effects of nutrients on bone loss—insights that can also be applied in clinical research (Shelhamer et al. 2020, see also Sect. 3.5.5).

3.5.3.3 Muscle Loss in Space and on Earth: New Knowledge on Signaling Pathways May Lead to New Countermeasure Targets

Muscle mass and strength are well-known to decline in response to microgravity exposure. However, despite the considerable knowledge gained on the physiological changes induced by spaceflight, the mechanisms of muscle atrophy and the effectiveness of inflight countermeasures still need to be fully elucidated. A full understanding of the mechanisms of disuse-muscle atrophy is even incomplete on Earth (Rennie et al. 2010). Nevertheless, a wealth of useful data has been obtained in space.

So, space life sciences research led to some new fundamental insights also regarding key signaling pathways in the neuromuscular system, e.g., for nitric oxide (NO) and HOMER protein. Here, comparative studies with mice on a BION satellite flight and on the ISS provided interesting information. While a long-lasting sensitivity to microgravity was demonstrated for the antigravity soleus muscle throughout a 3-month ISS mission, there was evidence for a possible resistance mechanism to microgravity-induced atrophy of the fast-twitch EDL (extensor digitorum longus) muscle. Moreover, and even more important, the results indicate that IGF-1 (insulin-like growth factor-1), NOS1 (nitric oxide synthase-1), and interleukin-6 might represent molecular targets for the development of countermeasures to prevent muscle wasting conditions on Earth (Sandonà et al. 2012; for review see also Blottner and Salanova 2015; Caiozzio and Baldwin 2020). A similar idea was followed successfully in a recent ISS experiment: By blocking the myostatin/activin A pathway in mice both muscle and bone loss was significantly inhibited (Lee et al. 2020b).

In addition to these new approaches, various classical countermeasures against muscle loss have been tested extensively in space and in bed rest studies (see Sect. 4.7). The use of combined physical (exercise), pharmacological, and nutritional interventions seem particularly promising (Pavy Le Traon et al. 2007). Vibration exercise studies on normal human skeletal muscle have shown benefits in sports (Mester et al. 2006; Cochrane 2011; Ritzmann et al. 2014), in sarcopenia in the elderly (Bogaerts et al. 2007), but also in various clinical settings such as for patients with neurodegenerative diseases (Sitja-Rabert et al. 2012) and in Duchenne muscular dystrophy (Soderpalm et al. 2013). These are examples of how findings from fundamental basic research in space life sciences may be translated to practical applications to mitigate or even prevent disuse-atrophy of apparently healthy populations—a great challenge for our modern sedentary society on Earth.

3.5.3.4 Cartilage: A Neglected Tissue in Musculoskeletal Research

In contrast to the substantial research on the degradation of muscles and bones in astronauts, cartilage unloading has long been neglected in musculoskeletal research.

Recent bed rest studies as well as a project on the ISS have now taken care of this important issue. For humans on Earth, the mechanical loads imposed by gravity and physical activity are important for preserving the function of the musculoskeletal system including that of cartilage, tendons, and ligaments. Moderate exercise is indispensable for a joint to supply its cartilage with nutrients and preserve its function. Any reduction in mechanical loads, for example, experienced by immobilized patients on Earth or caused by weightlessness in space, is likely to harm this tissue. As a bradytrophic tissue with a slow metabolism and limited capacity for regeneration, cartilage is extremely sensitive to disuse. Degenerated articular cartilage cannot recover again. Once it has been destroyed, an artificial joint is the only therapy option. The question is: To what extent is the articular cartilage affected negatively by mechanical unloading in space?

The ISS experiment "Cartilage" was designed by scientists from the German Sport University (DSHS) in Cologne to investigate the influence of microgravity exposure on the structure and biology of the knee cartilage in astronauts and to establish whether there is a connection with muscular atrophy. Knee tissues of astronauts were examined by magnetic resonance tomography (MRT) before and after flight to the ISS; also, volume and thickness of the cartilage were recorded. In addition, certain biomarkers of the cartilage metabolism were analyzed in blood and urine samples.

First, results of the experiment performed recently on the ISS indicate that the cartilage extracellular matrix is sensitive to prolonged exposure to microgravity, which is reflected by changes in serum molecular biomarker levels of cartilage turnover (Niehoff et al. 2016). The results of this first study to investigate the effect of microgravity on cartilage in space are supported by those from recent bed rest studies and studies under artificial gravity (Liphardt et al. 2018; Dreiner et al. 2020). Microgravity, bed rest studies, and studies on short-arm human centrifuges thus represent valuable models of unloading and will help to better understand the impact of immobility on cartilage health also in the context of disease or lifestyle on Earth. Future research will investigate the effects of exercise countermeasures on cartilage metabolism during spaceflight with the goal of minimizing deleterious effects of immobilization on cartilage health.

3.5.4   Immune Defense Problems in Space and in Intensive Care on Earth

Immunological problems of spaceflight were already discovered on the first Apollo missions in the late 1960s, where more than half of the astronauts suffered from bacterial or viral infections or showed inflammation-related symptoms (Mermel 2013). In Skylab and Soyuz crewmembers, a reduced reactivity of blood lymphoid cells was observed, indicating a potentially alarming observation for long-term space missions. About five decades of investigations have been conducted using biological samples like blood, urine, and saliva from astronauts. Also, studies with cell culture systems and animal models have been performed—inflight as well as in simulation facilities on ground. A wealth of information has been collected, clearly indicating that spaceflight affects the immune system however the magnitude and clinical significance of these alterations are still under discussion. The same is largely true for the causes and mechanisms underlying these impairments, the reason not the least being the enormous complexity of the immune system (Choukèr and Ullrich 2016; Choukèr 2020). Another limitation is that most of the data have been collected on short-term missions and via pre and postflight analyses only. A recent joint Russian/German study with pre, post, and inflight measurements on ISS astronauts living in space for more than 140 days sheds new light on these issues. Before going into more details some basic features of the immune system should be repeated first.

3.5.4.1 Elements of the Immune System: From Physical Barriers to Tailored Immune Responses

Three lines of defense have been established during evolution of higher vertebrates up to humans. The first line of host defense is of physical nature and consists of barriers like the skin as well as mucus, enzymes, and acidified thin layers of liquid. The skin, whose tallow, sweat, and bacteria act as microbial growth inhibitors, additionally regulates our water and temperature balance and protects us against UV radiation. According to NASA studies, skin problems such as dryness, scaliness, or itching are the third most frequent health complaint in astronauts, following headaches and vertigo. Delayed wound healing and increased allergic reactions to certain materials also figure among the problems reported. However, no systematic research efforts have been made so far to explore these disorders. Recently, scientists at Witten/Herdecke University studied the influence of space conditions on the physiology of the human skin. The ISS experiment involved a noninvasive examination of the skin on the inside of the lower arm, using three small measuring devices adapted for use in space: a device called corneometer that records the hydration level of the skin, a tewameter to measure trans-epidermal water loss, and a camera called visioscan to record images of the skin surface. Additionally, the capillary capacity of the skin (microcirculation), and the skin's ultrastructure and elasticity were measured immediately before and after the flight. The ISS experiment was performed between 2013 and 2018. To some surprise, the results showed no deterioration of the skin, but rather an improvement in skin hydration and skin barrier function. Also, skin density and thickness, as well as skin elasticity values, were unchanged from preflight data. In conclusion, no negative impact on skin physiological parameters was detected in space (Braun et al. 2019a).

To clarify the situation a questionnaire study was subsequently performed with 46 ISS crewmembers. The results of this examination clearly demonstrated that skin symptoms correlate with poor hygiene on orbit, whereas the specific environmental factors of spaceflight including microgravity obviously play a minor role. Based on these data, more attention to skin hygiene and maintenance should be recommended to the astronauts (Braun et al. 2019b).

3.5.4.2 Innate and Adaptive Immune Systems and Their Changes in Space

The immune system as such is a complex network of various organs, cell types, and molecules whose primary function is to distinguish between "own" and "foreign" thus being able to defend our body against pathogens. It helps with the healing of tissues damaged by trivial injuries or by surgery and destroys defective body cells. Formed in an early phase of evolution, the innate immune system can even be found in simple organisms, putting up an effective defense against pathogens. Only higher vertebrates including humans have developed a more sophisticated, adaptive immune defense system that protects us against pathogens even better. This response is targeted exactly to the typical pattern of a pathogen, also allowing the recognition of the microorganism for a further, more rapid and more efficient action in the course of a later, second contact. This "third-line" response of the adaptive immune system is mainly based on the action of B and T lymphocytes (Choukèr 2020; Choukèr and Ullrich 2016).

In space, special environmental conditions, such as microgravity and radiation, weaken the immune system of astronauts. In addition, stressors like isolation, a heavy workload, and a disturbed circadian rhythm, also contribute towards this impairment. Critically ill people on Earth struggle with comparable problems. Also, the risk of space missions for astronauts and the mortal danger threatening intensive care patients are—at least to some extent—comparable challenges. In these situations, the immune system should be strong enough to protect it from pathogens however the defense system should not be overburdened. Causes and mechanisms of impairments are still largely in the dark, which makes prevention rather difficult. Consequently, immune system research is important not only for astronauts but also for people on Earth. Space research offers excellent opportunities to investigate the influence of changing gravity conditions and other stressors. Therefore, all space agencies are active for many years to unravel the secrets of the malfunctioning immune system in astronauts. The final aim of all these projects is to elucidate the complex network of immune reactions with innovative methods—from the molecular and cellular level to the entire human organism—and then, in the next step, to develop effective countermeasures (for recent overviews see Buchheim et al. 2020; Crucian et al. 2018). Also, the German space life sciences program has defined immune system research as one of its priorities some years ago. Investigations on the overall immune system of astronauts as well as studies on tissues and cells on the molecular level are being performed.

For several years, scientists from Munich's Ludwig-Maximilians-University—in collaboration with American, Russian, and European scientists—have been conducting elaborate biochemical analyses complemented by psychological tests to study alterations in the immune systems of long-term ISS crew members. In conjunction with the results of isolation and bed rest studies, parabolic flights and medical records of individuals working at extreme locations such as the Antarctic, valuable information has been collected on the role played by individual factors that weaken the immune system. At the same time, the data provide new insights into the underlying mechanism of the human immune defense system. This knowledge is a necessary step on the way towards developing new preventive and therapeutic measures for the treatment of astronauts and intensive care patients with a severe illness. Currently, available countermeasures have recently been thoroughly summarized (Crucian et al. 2018).

The respective research project on the ISS began already in 2004 and was finalized in 2013. Initial results show that in astronauts on long-term missions, the so-called stress response system is activated. This includes the endocannabinoid system (ECS) that has so far hardly been investigated in space. However, it was known from ground studies that the ECS exerts a strong suppression of innate and adaptive immune responses (Buchheim et al. 2018). In fact, a constant exposure to a stressful environment and an ongoing inflammatory response can result in a pathogenic phenotype showing features of premature immune aging. This so-called inflammaging is a low-grade, pro-inflammatory state that renders its host at risk for latent viral infections, allergic, or autoimmune disease as seen in elderly patients. Indications that space travel might promote such an age-related inflammatory phenotype prematurely in astronauts had already been discussed. Indeed, the data from the physiological measurements together with the functional tests suggest a special conglomerate of stressors inflight, triggering a hyperinflammatory and aging immune phenotype (inflammaging) that persists even after 1 month after return to Earth thereby enhancing the risks of astronauts for hypersensitivity diseases such as allergies or autoimmune diseases. Further onboard testing and longer postflight observation periods are necessary to understand the underlying mechanisms and the detailed consequences for astronaut health. In the end, corresponding mitigation strategies based probably on a personalized approach should be developed (Buchheim et al. 2019).

3.5.4.3   From Space to the Patient: A New Assay to Monitor the Immune Status in Patients and Astronauts

Monitoring the immune status of humans under extreme environmental conditions such as spaceflight or in patients affected with human immunodeficiency virus (HIV) or in sepsis is of critical importance regarding the timing of adequate therapeutic countermeasures (for review see Choukèr 2020). For many years, the so-called in vivo skin delayed-type hypersensitivity test (DTH) served as a tool for this purpose. This test however was phased out in 2002, which brought scientists from the LMU Munich in Großhadern, Germany, to look for an alternative assay. This seemed especially necessary for their patients in the clinic, but also because the researchers were in the process of preparing a project on immune system research on the International Space Station ISS at that time. For this purpose, they tested a new alternative assay using elements of the skin DTH which is based on an ex vivo cytokine release from whole blood (for details see Feuerecker et al. 2012). For application in space, the assay was successfully tested on parabolic flights under acute stress as well as on the ISS under chronic stress conditions. Later, the assay was applied in HIV-infected patients and in patients with septic shock (Feuerecker et al. 2018; Kaufmann et al. 2013). In brief, the investigations showed the applicability of the new cytokine release assay developed originally for space application as a suitable diagnostic tool for immune monitoring of patients in the clinic. Further studies in larger patient cohorts are underway to determine the value of this cytokine release assay also as a tool for predicting the long-term survival of patients. First, results indeed confirm that the in vitro DTH assay may serve as a tool also for longitudinal follow-up of immune responses (Feuerecker et al. 2019).

3.5.4.4 Breath Gas Analysis for Noninvasively Monitoring the Immune Status

In another new approach, the status of an astronaut's immune system is monitored by means of a noninvasive breathing gas analysis thereby complementing blood tests and testing its potential to even replace them altogether in the future. The routine practice on the ISS is to freeze blood and urine samples and carry them back to Earth after a few weeks or months to have them analyzed in a terrestrial laboratory—an approach, certainly not feasible for long-term exploratory missions or stations on Moon or Mars. By this breath gas analysis approach, certain volatile substances in exhaled air are detected finally leading to online monitoring of the status of the immune system or illness-related changes in the human organism (Dolch et al. 2017, for details see Sect. 4.2.5).

3.5.4.5 Cellular Effects of Microgravity on the Innate and Adaptive Immune Systems

In addition to physiological research dealing with human beings in an integrative approach, also human cells have been and are still being studied by many research teams to shed a light on the mechanisms that cause the impairment of the immune system in microgravity. Immune cells constantly patrol all tissues and all organs in our body to defend it against dangers like infiltration by germs. In the process, they cover enormous distances day after day. They do not simply float around freely in tissue but constantly have to jump barriers to move from the circulating blood into the surrounding tissue, for example. Moreover, different immune cell types frequently attach to one another to exchange important information and activate each other. All this is necessary to repel foreign bodies. The cellular effects of altered gravity on the innate as well as on the adaptive human immune systems have recently been summarized, also providing overview tables on the experiments performed in space (Tauber and Ullrich 2016; Hauschild et al. 2016).

A possible cause for the impairment of immune system function may thus be an interference of microgravity with the movement or mutual attachment of immune cells. To pursue this question, scientists investigated the molecules and mechanisms which enable immune system cells to attach to one another, to migrate, and to break through tissue barriers. Results of projects run by the universities of Magdeburg and Zurich on various flight opportunities show that different cell types of the immune system react to the onset of microgravity in many different ways. Thus, human endothelial cells step up the production of ICAM-1, a surface molecule that is important for cell attachment. Conversely, monocytic cells react by reducing the production of ICAM-1. Since these cells, which are part of our innate immune defense system, attack exogenous intruders, microgravity interferes with an important mechanism that is indispensable for the monocytes' ability to attach themselves to surfaces and to other immune cells (Paulsen et al. 2015).

This finding might explain why specific immune cells no longer function properly in microgravity. These so-called T lymphocytes must remain attached closely and permanently to the monocytic cells so that they can divide and become completely activated—a process that does not work in microgravity. On the other hand, it would certainly be too simple to assume that these changes constitute the sole explanation for the astronauts' immune system weakness. More recent results from parabolic flights show that the so-called IL-2 receptor at the surface of T lymphocytes is toned down as well. This receptor mediates the signal that is essential for the reproduction of the activated lymphocytes. T cell regulation in microgravity has been reviewed nicely also providing a schematic summary of the available knowledge (Hauschild et al. 2014).

A recent experiment in the BIOLAB facility of the Columbus ISS module allowed for the first time the direct measurement of a cellular function in real-time and on orbit, namely the oxidative burst reaction in mammalian macrophages. It was shown that these cells adapted to microgravity within seconds, which demonstrates direct evidence of cellular sensitivity to gravity. This indicates that mammalian macrophages are equipped with a highly efficient adaptation potential to the microgravity environment (Thiel et al. 2017).

Also, for the first time, live cell imaging of primary human macrophages in microgravity was performed in 2018 on a TEXUS sounding rocket flight using FLUMIAS (Fluorescence Microscopy Analysis System) newly developed in the frame of the German space life sciences program (see Sect. 4.5). The images clearly showed geometric cellular changes and rapid response and adaptation of the potential gravity-transducing cytoskeleton within 20 s of microgravity with an adaptation after 2–3 min (Thiel et al. 2019). Further live cell imaging studies with FLUMIAS also on the ISS are hoped to provide deeper insight into these changes in the future.

3.5.4.6 Fighting Cancer with Space Research

On a first glance, cancer and space research may seem to be a surprising context. However, microgravity, as well as simulation experiments on ground, have shown that altered gravity conditions provide a unique environment for studying and influencing tumor cell processes (for review see Becker and Souza 2013; Krüger et al. 2017). In fact, research on cancer biology in space started already during the Space Shuttle mission STS-90 in 1998. Primary cultures of human renal cortical cells showed remarkable alterations of genes after 6 days in microgravity. In the following years, experiments demonstrated that microgravity changes microtubules and mitochondria of cancer cells, modifies the production and structure of cytoskeletal and extracellular matrix proteins, induces apoptosis, and changes the secretome.

In a recent review, the results of 20 years of research in this area have been summarized (Krüger et al. 2019). Of special importance is the finding that adherently growing cancer cells detach from their surface and form three-dimensional structures in microgravity. The cells included in these multicellular spheroids (MCS) turned out to more closely mimicking the conditions in the human body thus becoming an invaluable model for studying metastasis and developing new cancer strategies via drug targeting. Microgravity intervenes deeply in processes such as apoptosis and in structural changes involving the cytoskeleton and the extracellular matrix, which influence cell growth. Most interestingly, follicular thyroid cancer cells grown under microgravity conditions were shifted towards a less malignant phenotype. Also, a large number of proteins which may serve as promising targets for drug development as well as available drugs have been identified. The results may truly provide hope for thyroid cancer patients.

As mentioned above, a new approach in cell research has recently started by applying the FLUMIAS microscope. By using MCF-7 breast cancer cells, live cell imaging with FLUMIAS during the TEXUS-54 sounding rocket mission visualized similar changes as those occurring in thyroid cancer cells in microgravity. This result indicates the presence of a common mechanism of gravity perception and sensation in cancer cells (Nassef et al. 2019).

Certainly, microgravity research is just one part of the cancer research performed in many labs around the globe. Nevertheless, these results from space research may be used to rethink cancer research on Earth with the aim of developing new drugs and cancer treatment strategies based on novel target proteins identified by microgravity research.

3.5.4.7   Tissue Engineering in Microgravity: New Approaches in Regenerative Medicine

Tissue engineering enables the development of functional constructs from cells with application potential in regenerative medicine and drug screening as well as in nontherapeutic approaches. Early experiments onboard of several space shuttle missions already in the 1980s had revealed that microgravity might be beneficial for the formation of three-dimensional cell aggregates and led to further attempts to study this phenomenon more thoroughly (for review see Wehland and Grimm 2017).

Since isolated cells cultivated in culture flasks will only grow in a 2D monolayer on Earth, a so-called scaffold is introduced that helps the cells to assemble in a 3D structure. Since scaffolds however might eventually lead to immunological problems, in the long run, the ultimate goal in tissue engineering remains the de novo formation of scaffold-free, functional, and organotypic tissue constructs.

To achieve this goal a lot of experiments have been performed in microgravity as well as in devices simulating weightlessness on Earth. Chondrocytes for cartilage generation, thyroid cancer cells to produce spheroids for further studies of tumor development, metastasis, host-tumor interaction, and drug testing as well as bone tissue engineering from marrow stem cells, osteoblasts, or mesenchymal stem cells were tested. The results are promising, but the implication of all this must be clarified in further experiments. Especially, the underlying mechanisms of the observed spontaneous cell aggregation are not fully understood.

Recently, it was shown for the first time in an ISS project that endothelial cells exhibit three-dimensional growth forming tube-like cell aggregates resembling the intima of small, rudimentary blood vessels (Grimm et al. 2017). Since they grow as single layers on the bottom surface of cell culture flasks under normal (1 g) culture conditions, the results clearly indicate that microgravity is the major cause for the formation of tubulin-like aggregates. Also, IL-6 is more and more emerging as a key protein in the microgravity-dependent formation of 3D-growth, since it is up-regulated, when 3D structures are formed.

This knowledge and further results from ongoing space experiments contribute to an optimization of the three-dimensional tissue growth and the understanding of scaffold-free tissue engineering. With further investigation of the space samples, scientists hope to understand endothelial 3D growth and to improve the in vitro engineering of biocompatible vessels which could be used in surgery as medical transplants (Grimm et al. 2017). Additionally, these constructs provide an efficient tool for downstream experiments such a drug testing and could be used as a replacement for in vivo models thereby reducing the need for animal testing. One aspect should be stressed clearly to avoid any misunderstanding: The goal of these projects is not to run tissue engineering experiments in space for production purposes but to learn from the space experiments for enabling or improving tissue engineering on Earth.

3.5.5 A New Concept for the Role of Salt for the Immune System, Bone, and High Blood Pressure

One of the finest examples in space life sciences for the importance of the already emphasized integrative view on human physiology and human health is the elucidation of the role of salt in nutrition. Respective experiments started by preparatory terrestrial isolation and confinement studies initiated by ESA, DLR, and IBMP Moscow in the early 1990s followed by space projects on the German-Russian MIR'92 and the Shuttle/Spacelab D-2 missions in 1993 (for review see Gerzer 2014). Continued during the MIR'97 mission and the Russian isolation study

Mars500 research was complemented by experiments on the ISS and further clinical studies during the last years. In these studies, scientists from the University Clinic of Erlangen, the University of Bonn, and from the DLR Institute of Aerospace Medicine in Cologne—together with their Russian and American cooperation partners—established a totally new view on the context of salt and various physiological systems of the human body leading to the provocative headline in the New York Times "Why everything we know about salt maybe wrong" in May 2017 (www.nytimes.com/2017/05/08/health/salt-health-effects.html).

Sodium is the major cation of the extracellular fluid. Together with chloride, it is utilized by the human body to maintain normal water distribution, osmotic pressure, and anion-cation balance. In fact, without sodium chloride humans would be unable to survive: the transport of water and nutrients in our bodies would cease, nerves would have difficulties transmitting signals, and the muscles, too, would be unable to function properly. But there is a reservation: too much salt has a negative effect on many of our body systems and processes such as blood pressure regulation and bone metabolism (Heer et al. 2015).

Textbook knowledge has stated for decades—basically since early experiments in the middle of the nineteenth century—that there is a strong coupling of salt and water. A healthy human body tries to keep its salt concentration stable. After ingesting too much salt, the body takes certain countermeasures to rid itself of surplus sodium. The first reaction is a feeling of thirst leading to drinking and thus to accumulation of more water in the tissue to keep the salt concentration in the human body fluids constant. The next step is to eliminate more sodium and water via the kidneys—so much in brief for the established dogma.

Results from the German D-2 and MIR'92 space missions and accompanying simulation studies challenged this dogma already in the early 1990s (Drummer et al. 2000; Heer et al. 2000). Metabolic balance studies on ground followed indeed showing a neutral balance of water, while there was a high storage of sodium leading to the assumption of an additional—until hitherto unknown—sodium storage mechanism. In studies on rats this additional location for storage of sodium was found—it is the skin that acts as a potent storage location for sodium. It is bound to glycosaminoglycans in exchange with protons. Close examination of the rat skin also revealed a macrophage migration into the skin as well as an increased formation of lymph vessels (Titze et al. 2005).

Besides defending the body against infections, macrophages appear also to be involved in the regulation of the salt balance. In body regions with high salt concentrations, they cause the formation of new blood and lymph vessels in the skin thus helping to regulate the body's microcirculation. The latter, in turn, is also connected with blood pressure regulation so that macrophages ultimately play a part in controlling blood pressure as well (Machnik et al. 2009).

These findings are also relevant from the clinical point of view, in that they explain why some patients do respond to a high salt intake with a rise in blood pressure, and others do not. Contrary to previous assumptions, blood pressure variations after increased salt intake are, in fact, caused by the salt itself, rather than by the water accumulation following a high salt intake—a new treatment perspective for hypertension.

The observation that sodium is exchanged with protons also suggested that high sodium ingestion leads to acidosis which in turn is known to activate osteoclasts and to stimulate bone and protein degradation. Indeed, markers of bone degradation were found to be strongly increased when a high sodium diet was given, and protein degradation was also found to be elevated. Both bone and protein degradation can be counteracted by the addition of bicarbonate to the diet, which is of high potential clinical relevance since bone as well as muscle degradation are important problems not only for astronauts during their stay in space but also for people on Earth during the aging process. Introducing meals with low sodium and alkaline pH might thus be greatly beneficial in keeping human beings healthy (Frings-Meuthen et al. 2008).

This concept was tested in ISS investigations between 2008 and 2012. Experts from the DLR Institute of Aerospace Medicine and from Bonn University have studied the interaction between nutrition, salt, fluids, acid/alkaline balance, and human bone metabolism. For this purpose, astronauts had to spend two 5-day periods on the ISS following a defined diet, one period with a high salt content and one with a low one. Their excretion of salt and that of special marker substances which deliver indications as to their bone metabolism was recorded in the urine. Blood tests provided additional evidence on further bone markers and the hormones involved in the salt/water balance. Regular weight checks provided information to the scientists on how much fluid has been given off through the astronauts' skin. Results indicated that during periods with a high salt intake the excretion of calcium is higher than during the 5-day period with a low salt intake. This, in the long run, might lead to a negative calcium balance, higher bone resorption, and a higher risk of developing kidney stones. Taken together the data are interpreted as an indication of a resetting of natriuretic peptide in astronauts (Frings-Meuthen et al. 2020). Applied to conditions on Earth, a low level of physical activity—as it is common in elderly and bedridden patients—combined with a high salt intake would thus lead to increased bone decay. A combination of dietary measures and exercise programs could offset this tendency and improve healthcare outcomes.

In the same timeframe, investigations were continued during the Mars500 isolation study. The study was carried out between June 2010 and November 2011 at the Moscow-based Institute of Biomedical Problems (IBMP) with six male subjects. Its purpose was to find out more about the effects of a long period of isolation on human physiological and mental conditions. In a closed "space capsule" it was possible to meticulously monitor food intake and excretion; the setup permitted a day-by-day comparison of each subject's prescribed dose of salt specified on their diet plan with sodium concentrations in the urine. According to the conventional dogma, the exact amount of sodium taken up with the food should have left the body again within 24 h, yet, despite the intake of identical amounts each day, there was an enormous variance in daily excretion. A more detailed analysis soon revealed that the excretion of salt follows a certain rhythm, with cycles lasting 7, 14, and 28 days, respectively. Also, two hormones, aldosterone and cortisol, which so far have been believed to have a similar effect, apparently act as antagonists in the accumulation and excretion of sodium, and their concentrations, too, vary in cycles—in other words, there seems to be a kind of "male period" (Rakova et al. 2013). Thus, it is not only the kidney that is responsible for the regulation of the salt-water balance, but certain hitherto uninvestigated hormonal factors play a part as well. The recent findings will also reshape hospital routines. We now know that collecting 24-h urine is not sufficient to assess a patient's salt balance reliably. And there was another assumption that the scientists from Erlangen could confirm during the Mars500 experiment: Ingesting less salt did, in fact, lower the subjects' blood pressure—even that of healthy subjects in Mars500 (see also Gerzer 2014).

Recently the results of Mars500 and further clinical studies have been summarized (Rakova et al. 2017; Kitada et al. 2017) with additionally provoking a commentary contribution in the same journal under the headline "Salt and water: not so simple" (Zeidel 2017). In summary, the publications demonstrate that osmotic balance in response to high salt intake involves a complex regulatory process that is influenced by hormone fluctuations, metabolism, food consumption, water intake, and renal salt and water excretion. Also, the immune system is involved in blood pressure regulation, and this—together with the finding of the human skin acting as an important storage location for sodium—implies that there is probably a close link between the new salt storage mechanism and hypertension. With these data, the relative role of the kidney, the roles of the interstitium, the skin, and the immune system all need to be reevaluated. Current therapeutic regimes need to be questioned as well, and new targets for drug development like the immune system or the regulatory mechanisms of the new salt storage mechanisms may arise.

All this is of high economic importance: The 12 grams of that humans consume every day in industrialized countries is twice the daily intake recommended by the World Health Organization. According to experts, an appropriate reduction of salt consumption in the USA would mean between 66,000 and 120,000 fewer cases of coronary heart disease, between 32,000 and 66,000 fewer strokes and between 54,000 and 99,000 fewer heart attacks—saving the health system between 10 and 24 billion dollars (Cappuccio 2013).

In addition to the fascinating scientific results described above, it should be pointed out that during these studies an innovative, noninvasive method for detecting sodium in the skin and other tissues was developed, ^{23}Na MRI (magnetic resonance imaging), and registered for patent a few years ago (Kopp et al. 2012). Meanwhile, the new method to detect sodium accumulation is successfully applied in the clinic, for instance in patients with hypertension, sclerosis, or diabetes (Kopp et al. 2013, 2017, 2018). Sodium storage was found to increase with age in muscles and skin, the increase being more pronounced in men than in women and is directly associated with blood pressure levels. Following the changes of tissue sodium storage achieved by lifestyle changes or medication may thus be a new conceptual approach (Kopp et al. 2013). Also, ^{23}Na MRI might be a promising tool to assess the sodium level in the skin for predicting the progression of skin fibrosis in patients suffering from systemic sclerosis (Kopp et al. 2017).

In summary (see also Gerzer 2014), results from internationally coordinated space experiments complemented by preparatory and accompanying terrestrial studies are revolutionizing established clinical concepts by demonstrating that

- The dogma of salt storage and the close correlation of sodium with water balance is wrong
- The human skin is an important storage tissue and regulator for sodium
- Sodium storage involves macrophages which in turn salt-dependently coregulate blood pressure
- Sodium also strongly influences bone and protein metabolism
- Immune functions are also strongly influenced by sodium
- During the aging process body sodium storage is increased—in men more than in women—which in turn might enhance the aging process of the human body.
- Taken together, the results lead to new approaches for counteracting diseases such as hypertension and osteoporosis
- ^{23}Na MRI was established as an innovative and noninvasive method to visualize sodium in the skin and other tissues also in patients.

This fascinating "salt story" bears several messages that scientists but also politicians and the public should consider when judging the benefits of their investments in (space) science and research: It can take many years before a breakthrough becomes evident and understood—in this case nearly 30 years. Space and ground-based research are complementary and need to go hand-in-hand. And last but not least, this story obviously proves the importance of integrative physiology that only space research has promoted over the years (see also Ruyters and Stang 2016).

3.5.6 Living Under Extreme Conditions: Human Health and Performance Issues

Living in space such as onboard of the ISS is living under extreme conditions. While the absence of gravity primarily challenges the physiological systems of astronauts as pointed out in several chapters of this book, we should not forget that spaceflight combines a variety of risk factors for physiological and mental health. It includes physical (e.g., weightlessness, radiation, missing sunlight, and day-night cycle), psychological (e.g., mission goals, duration, workload, sensory and sleep deprivation, stress), and social factors (e.g., crew dynamics, heterogeneous crews, loneliness) (Sandal et al. 2006; Rosnet et al. 2000). Thus, physiological health problems, as well as mental impairments, are likely to occur as demonstrated by space-related research of the past decades (Fowler and Manzey 2000; Manzey 2000; Demontis et al. 2017). Astronauts will increasingly face these problems during the upcoming long-duration exploratory missions.

Mental impairment due to a variety of stressors is, of course, not only a concern for astronauts, but also for people living on Earth. In everyday life, we are enclosed in different social groups and we need to develop strategies to cope with these groups and their individual members. Both, living in extreme environmental conditions such as space and living in the modern society, are accompanied by stressors classified as social (work-life balance), mental (high workload), and physical (sedentary lifestyle) stressors. In space and other extreme environments however changes occur much faster than in normal life representing a time-lapse of a sedentary lifestyle and aging. Accordingly, results obtained from space research or isolation studies may also contribute to present debates on active living, and its relevance for socioeconomic and health-political decisions in the near and far future.

This is especially true in the current situation, in which the world is experiencing in a certain sense the largest isolation experiment in history: In an attempt to slow down the spread of the COVID-19 pandemic numerous countries across the world have shut down economy, education, and public life; governments have decided on strict regulations of quarantine and of social distancing. The consequences of these measures on brain, behavior, neuro-humoral, and immunological responses are largely unknown. Since space research has a long history in seeking to understand the effects of isolation and confinement, the value of shared learnings from spaceflight during COVID-19 is obvious and has indeed been emphasized by scientists, in news reports and in social media channels (Choukèr and Stahn 2020).

In spaceflight and in analog studies, cognitive and neurophysiological changes have been demonstrated, the results being divergent, however (for review see Abeln et al. 2016 and references therein). While evidence for cognitive impairments concerning psychomotor speed, internal timekeeping, attentional processes, and central management of concurrent tasks exists, other studies fail to identify cognitive impairment or even show positive effects. Also, reports of disturbed decision-making and judging in space are available, which may have fatal consequences.

In addition, psychological reactions such as decreased attention, vigor, motivation, and appetite as well as higher emotional and perceptual sensitivity, fatigue, psychosomatic symptoms, and interpersonal conflicts were reported. Also, the existing results of long-term isolation on affective states during spaceflight and in analog conditions are inconsistent. While some crews show affective declines around mission midpoint or within the third quarter, some show declining emotions over mission duration, and others no emotional declines at all or even positive outcomes.

Moreover, the underlying neurophysiological processes remain widely unclear. Hemodynamic changes such as the well-known blood volume shifts to the upper body can probably be excluded to be the reason for cognitive decline in space (Klein et al. 2019, 2020). Endocrinological changes, especially in stress hormones are considered to play a role in the response and adaptation in physiological and psychological systems. Further investigations are necessary to determine the importance of endocrinological changes more clearly and their impact on cognitive and affective performance.

Due to methodological restrictions, studies examining electrocortical changes in space or analog environments are rare (Schneider et al. 2013; Roberts et al. 2020). Several studies lead to the conclusion that electrocortical activation does not relate to central hemodynamic shifts but might reflect neuro-affective and neurocognitive adaptations (for review see Abeln et al. 2016, and reference therein; Weber et al. 2019; Strangman et al. 2020). So far little attention has been given to long-term investigations of brain cortical activity in space (Cheron et al. 2006), which makes it difficult to draw any manifest conclusion. The implementation of new integrative, easily applicable, and usable neuroimaging technologies (EEG, fNIRS) in the International Space Station (ISS) is planned and should contribute to new insights into the effect of long-term exposure to weightlessness and its circumstances on the neurophysiological and neuropsychological state. In conclusion, it seems fair to say that mental health research in space is still in its infancy. In view of the planned exploratory missions, more emphasis on these health aspects is certainly mandatory.

3.5.6.1   Exercise for Mental Health in Space and on Earth: Recommendations for the Future

In the past decades, space research mainly focused on investigating the effects of stressors of spaceflight, especially microgravity towards physical health thereby emphasizing the importance of exercise for maintaining physical fitness. But what about exercise and mental health? On Earth, positive effects of a regular exercise regime on stress problems and burnout, even on the probability to develop Alzheimer's disease and other neurodegenerative diseases have been shown. Exercise has been proven to have a positive impact on social, mental, and physiological stressors (for review see Abeln et al. 2016).

In the frame of the German space life Sciences program, especially a scientific team from the German Sports University in Cologne has significantly contributed to solving the issue of exercise effects on mental health (Abeln et al. 2015). Benefits of exercise on mental health include the enhancement of cognitive performance as well as an improvement of affective state. Studies on exercise programs and their influence on mental health also indicate a substantial involvement of exercise-induced central neuronal adaptations (Ekkekakis and Acevedo 2006). In particular frontal brain regions that are associated to play an important role in information as well as emotional processing seem to be positively affected by exercise (Brümmer et al. 2011; Schneider et al. 2009). These modifications are supposed to act as a multifunctional generator for the adaptation of mood, vigilance, and cognitive performance. Accordingly, medical interest on embedding such beneficial exercise programs to enhance neurocognitive functioning as well as to improve neuro-affective processing is increasing. The aim is to counteract emerging pathological challenges of our aging society and to reduce the consequences of sedentary lifestyles such as neurological degenerations.

Within space research, exercise is constantly used as a countermeasure against cardiovascular and musculoskeletal deconditioning for decades. So far, the implication of exercise for mental health has been neglected also due to the relatively young and deficient state of literature in this respect. However, applying exercise not only for physical but also for mental health has gained interest within the past years as a consequence of the abovementioned positive outcomes of exercise interventions on Earth and recent space-related investigations (Abeln et al. 2015; Basner et al. 2013; Schneider et al. 2010, 2013). Basner and colleagues just recently reported a correlation between decreased physical activity levels, sleep quality and quantity as well as vigilance during the Mars500 project.

Which exercise programs are necessary for the benefit of mental health, in particular, when facing the restrictive circumstances as present during spaceflights? First, it must be mentioned that exercise mode, duration, and intensity seem to matter. Especially running exercise causes a higher state of cortical relaxation within frontal brain regions, which are associated with higher executive functions (Brümmer et al. 2011; Schneider et al. 2009). Furthermore, the consideration of individual exercise preferences and freely chosen instead of prescribed exercise protocols seem to have an important impact (Ekkekakis 2009). There is also space-related evidence for this individual preference hypothesis: Schneider et al. (2013) were able to show that running exercise resulted in frontal cortex deactivation (relaxation) and higher cognitive performance during long-term isolation. Interestingly, the crew indicated running to be the most preferred exercise. Motivation, commitment, and compliance for exercise are obviously higher when individual preferences are appreciated. These aspects should not be underestimated when we aim for regular exercise to obtain physical and mental health at longer duration space missions.

This individualization of exercise programs might also be of relevance for Earthbound therapies using exercise programs to avoid a sedentary lifestyle with all its negative effects including physical and mental degeneration. Technological developments that are supportive to the individual's motivation to exercise—both, on Earth and in space—might be useful, such as the implementation of virtual realities, allowing for jogging or cycling in familiar environments like a forest home track. In conclusion, living in space, living under extreme conditions provides us with a time-lapse of deconditioning of the human physiological and psychological systems and offers the opportunity to develop, foresee, and counteract the degeneration in our aging civilized world and to stress the importance of exercise on physical and mental health.

3.5.6.2   Healthlab: Standardized Psychophysiological Measurements of Stress Effects

Many studies have shown negative effects of stress on mental health and cognitive performance. However, for a long time, it has not been possible to verify these experiences with measurable data using a standardized experimental design, neither in space nor on Earth. An additional problem comes with the observation that the autonomous nervous system, i.e., that part of the nervous system that cannot be consciously controlled, reacts to psychological burdens such as stress differently in different people. In many people, this translates to an increase in blood pressure,

Fig. 3.7 The Russian cosmonaut Oleg Kononenko used the Healthlab hardware on ISS to measure sets of psychophysiological data. (© DLR/NASA)

some break out in a sweat, while others feel gastric heaviness. These effects are known as the "reaction pattern of the autonomous nervous system."

Already in the late 1970s, efforts were undertaken especially in the Soviet space program to overcome this unsatisfactory situation. Since 1996, the so-called Pilot experiment has been implemented as a cooperative Russian/German space project aiming to investigate the cosmonaut's skills by performing a simulated manual docking of a Soyuz spacecraft to the space station, which is considered to be an important way to analyze the crew's operational reliability. In parallel to assessing the success of the docking, a suite of physiological and psychological parameters is measured among them ECG, respiration, blood pressure pulse transition times, skin conductance, EMG, and voice pitch. Thus, for the first time, psychological test procedures were combined with quantitative physiological measurements (Johannes et al. 2003). For these measurements, a special equipment called "Healthlab" was developed and consequently upgraded over the years in the frame of DLR's Space life sciences program (for details of the methodology see Johannes et al. 2003, 2008). The joint Russian-German project started on the MIR space station, where it was performed between 1996 and 2001. Again, in collaboration with IBMP, the second phase of the project was implemented on the Russian segment of ISS between 2008 and 2011 (Johannes et al. 2016a) (Fig. 3.7).

Overall, the PILOT experiment demonstrated that the performance level of Russian cosmonauts in a mission-relevant maneuver, namely the hand-controlled docking of a spacecraft, was found to be significantly improved on the ISS in comparison to the MIR station the main reason probably being the increased number of training sessions. Indeed, findings from the MIR station had suggested that a break in docking training of about 90 days significantly degraded performance (Johannes et al. 2016a).

The third phase of the project on ISS has started in 2015, the equipment being upgraded with the capability of also measuring EEG signals. Using source localization algorithms, it is aimed to assess mental load and cognitive performance by using neurocognitive markers (P300) and to correlate these with docking performance. The devices were successfully tested on three parabolic flight campaigns in 2015 and then launched to the ISS. In addition to Russian cosmonauts, US astronauts are also foreseen to participate.

Interestingly, Healthlab was also applied in various space-related isolation and bed rest studies in addition to its utilization in the Russian module of ISS. The German military forces used it for the objective psychophysiological strain assessment of AWACS (Airborne Warning and Control System) crews in simulators and real flight operations (Ledderhos et al. 2010). On the island of Curacao, Healthlab—especially using its in-water EEG measurement capability—was successfully tested for objectively measuring the efficacy of the so-called Dolphin Assisted Therapy (Johannes et al. 2016b).

3.5.6.3   Wireless Monitoring of Changes in Crew Relations During a Long-Duration Mission Simulation (Mars500)

Another field of application of Healthlab is the wireless monitoring of crew relations. During exploratory human missions, astronauts are exposed to long periods of autonomy, isolation, and confinement. It is thus of the highest importance to reliably measure crew performance and individual well-being in order to safeguard the success of the missions. Crew cohesion and its dynamics play a central role when coping with extreme physical, social, and psychological conditions with reduced communication, periods of high workload, and periods of monotony and boredom. Friendly and successful operational interactions of all crew members are known to be particularly important under those conditions; however, often interpersonal problems among crew members of isolated and confined groups are inevitable (Kanas 2020).

To cope with these situations, several approaches for monitoring the interpersonal relationship and crew cohesion have been taken over the last decades. Based on the experience with Healthlab and the Pilot space project a new system was developed and tested during the long-term isolation study Mars500 in Moscow. The goal was not only to measure certain parameters but also to develop a system that can provide objective feedback on crew cohesion to the crew itself. The results from Mars500 convincingly show the validity of the wireless group structure monitoring system for measuring cohesion when taking behavioral and activity patterns into account (Johannes et al. 2015). Future application on the ISS or on the upcoming exploratory missions seems reasonable.

3.5.6.4   Somnowatch: Sleep Monitoring via Simple Assessment of the Autonomic Nervous System

Astronauts often complain about sleep disorders or disturbances of their circadian rhythms, which is identified as risk for exploration missions. In fact, adequate sleep quantity and quality are required on Earth as well as in space to maintain vigilance, cognitive, and learning processes (Barger et al. 2020). Recent countermeasures have been implemented on the ISS to better regulate sleep opportunities and to allocate enough time for sleep. However, for proper sleep recovery, the quality of sleep is also critical. Most related ISS studies focused on sleep quantity and chronobiology since during human spaceflight all-night sleep recording is hard to achieve (for review see Clément and Ngo-Anh 2013). New approaches and new developments have therefore been initiated recently (Penzel et al. 2016; Petit et al. 2019).

The approach of the research team in Berlin focuses on the idea to use the noninvasive assessment of cardiac autonomic nervous system (ANS) with heart rate variability (HRV) analysis which can be achieved by rather simple sensor technology (Penzel et al. 2016). Since the ANS is modulated by sleep and wakefulness, monitoring its malfunctioning due to sleep deprivation provides a window for tracking sleep disorders noninvasively and without significant disturbance to astronauts as well as of subjects on Earth. Different devices such as Somnowatch—currently being used to measure blood pressure within the spacesuit of astronauts—or the advanced version "Somnotouch RESP" (www.somnomedics.de) have been successfully applied during isolation studies such as Mars500 and SIRIUS (Scientific International Research in a Unique terrestrial Station), the latter performed at IBMP Moscow as well as in the hospital on patients with sleep apnea (Fietze et al. 2015). A validation study comparing data from these devices with the gold standard method polysomnography had shown that reliable data such as total sleep time, sleep period time, sleep efficiency, sustained sleep efficiency, sleep onset latency, and sleep-wake pattern are obtained (Dick et al. 2010). Therefore, this new easy and quick method is highly suitable to assess sleep patterns in astronauts and in patients on Earth suffering for instance from sleep apnea as well as for occupational health issues, e.g., of pilots and truck drivers (Fietze et al. 2015).

3.5.7 The Twins Study and Long-Duration Missions

Extended duration missions on the ISS are steppingstones for future missions to the Moon and Mars. As a part of the preparation for these future destinations of human spaceflight, NASA started in 2015 to send astronauts on long-duration missions of approx. 1 year to the ISS. The goal of extended duration missions is, among other things, to better understand the impact of isolation, confinement, and environmental factors on human health and mission performance as well as to demonstrate and verify the techniques needed to prevent, recognize, treat, mitigate, or cure any negative psychological and psychophysiological effects of prolonged isolation and confinement during spaceflight (https://www.nasa.gov/1ym/about).

The first and at the same time extremely ambitious and highly demanding project in this context was called "Twins Study" with the objective to investigate health impacts of long-duration spaceflight on the twin US astronauts Mark and Scott Kelly. Scott stayed onboard of the ISS for 340 days, whereas Mark remained on ground (https://www.nasa.gov/hrp/one-year-mission-and-twins). Both twins underwent the same numerous preflight, inflight, and postflight examinations. The

Fig. 3.8 The Twins Study examined US astronaut Scott Kelly during a year in space and his twin brother Mark Kelly, also a former NASA astronaut, on Earth during the same period of time to investigate the long-term effects of microgravity and other space conditions on the human body. (© NASA)

multidisciplinary, multi-omics, molecular, physiological, and behavioral study is an extraordinary example for an integrative research approach under perfectly controlled conditions on ISS and on ground using two almost identical human beings. A first summary of the wealth of findings of the comprehensive investigation was recently published (Garrett-Bakelman et al. 2019) (Fig. 3.8).

Although the Twins Study was led by NASA, Russian cosmonaut Mikhail Kornienko co-participated in the long-duration mission. He collaborated with Scott Kelly in multiple investigations and added to the database that was being established. After the successful Twins Study, NASA continued with three further long-duration missions. During the years 2017–2020, NASA astronauts Whitson, Morgan, and Koch stayed onboard of ISS for extended periods of time between 272 and 328 days. Each of these missions helped defining the so-called Spaceflight Standard Measures, a uniform approach to measure the effects of spaceflight on the human body, which are to be implemented in the new long-duration mission projects NASA is preparing as a series for the next years.

3.5.8   Geriatrics Meets Spaceflight: Space, Gravity, and Physiology of Aging

"Falls and Fall-Prevention in Older Persons: Geriatrics Meets Spaceflight" is the somewhat provocative title of a recent publication by Goswami (2017), in which the author summarizes and compares spaceflight-induced physiological changes in astronauts with those occurring during the aging process of people on Earth. Even more detailed information on this topic is provided by a special volume of the journal "Frontiers in Physiology" (Goswami et al. 2020). In fact, the similarities are striking with spaceflight-induced changes in physiology representing a time-lapse of the aging process on Earth.

Only with the advent of human spaceflight, it became possible to explore the role of gravity in human physiology. What was totally unexpected at that time was a possible link between gravity and aging. The results especially of the Skylab missions in the 1970s demonstrated a variety of changes such as bone density loss, negative calcium balance, muscle atrophy, cardiovascular and hematological changes, metabolic, endocrine, and sleep problems so that medical observers commented that astronauts must be growing old faster. This conclusion was quickly dismissed when it became clear that the astronauts would recover rather soon after the missions (today we know that this is not entirely the case).

Today, it is generally accepted that the observed spaceflight-induced physiological changes as well as those in bed rest studies share important common features with the deconditioning and the impaired functions due to the aging process. Probably the first review on this topic was published by Vernikos and Schneider in 2010 based on earlier work focusing on the effects of inactivity. Reviews on the results of bed rest studies were presented earlier by Sandler and Vernikos (1986) and Pavy Le Traon et al. (2007). Most recently, Vernikos (2020) again compared changes induced by space or bed rest with those of aging, convincingly listing similarities in the changes as well as in the mechanisms involved. In principle, the physiological responses to space as well as during aging are characterized to a large extent by problems of mechano-transduction. In space, the weight of the human body is nullified due to the microgravity conditions thus resulting in deconditioning from gravity deprivation. On Earth, similar changes in deconditioning result from gravity withdrawal due to a sedentary lifestyle or by being hospitalized—in the presence of gravity. In both cases, there is a downregulation of mechano-transduction. Ingber's tensegrity model provides the framework for understanding how external and internal mechanical forces influence biological control at the molecular and cellular levels. His work reveals that molecules, cells, tissues, organs, and our entire body use tensegrity architecture to mechanistically stabilize their shape and integrate structure and function at all scales (Ingber 2008).

All in all, space medical doctors providing care to astronauts have much to learn from the study of the physiology of aging. Similarly, gerontologists, rehabilitation specialists, clinicians, and medical doctors on Earth, in general, can benefit from stepping back to appreciate the results of research and developments in space and to consider the physiological changes of aging primarily as a disorder of mechano-transductions. Medical and surgical advances have contributed to the extension of life. However, technological inventions encourage an increasingly sedentary lifestyle leading to deconditioning with all its accelerated physical incapacitations. In this context, gravitational physiology may lead to useful insights to help delay or prevent this physical incapacitation that seems inevitable with living longer. The passage of time is inevitable—improved quality of life is the goal enabled also due to the results of life sciences research in space (Vernikos and Schneider 2010; Vernikos 2020).

3.6 Summary of Key Findings and Breakthroughs in Space Life Sciences

Based on the accomplishments described above, a comprehensive summary of key findings and breakthroughs from 40 years of space life sciences research is presented in the following. While the focus here is certainly on European achievements, especially those achieved within the framework of the German space life science program, international contributions are also recognized as can be seen from the extensive list of references at the end of this chapter.

3.6.1 Space Biology and Bioregenerative Life Support

Gravitational Biology

- Complete elucidation of the signal transduction chain for gravitaxis in single motile cells, especially in Euglena, thereby settling a 100-year-old argument: orientation is based on active gravity-sensing.
- Complete elucidation of the signal transduction chain for gravitropism in single-celled model systems (green alga Chara).
- Significant progress in the elucidation of the complex signal transduction chain for gravitropism in higher plants, e.g., in Arabidopsis; transfer from morphological observations and physiological analysis to the molecular basis of gravity-sensing and graviorientation.
- Elucidation of the interaction between light- and gravity-induced growth responses in higher plants.
- Altered gravity affects the structure and function of animal gravisensory systems and their neural responses leading to adaptation and altered postural control and spatial orientation.

Bioregenerative Life Support Systems

- Plants can grow seed-to-seed in space and can be cultivated in proper cultivation systems to support human exploration missions.
- Significant progress in the understanding of the complex interactions between the various biological components and in the technical design with special emphasis on aquatic systems.

Astrobiology

- New concept for the formation of the planets of the solar system reshaping our understanding of the face of the primitive Earth.
- Identification of life in the form of microfossils or life-indicative chemical imbalances in the oldest rocks of our planet giving rise to the idea that Earth was inhabited much earlier than previously imagined, possibly leading to new concepts for the pathway of evolution from chemistry towards biological molecules and systems.
- Strong indications for the remarkable potential of life for adaptation to and survival in environments formerly considered too extreme to harbor life.

Radiation Biology

- Beyond the direct benefit for radiation protection of astronauts, space research focusing on the elucidation of radiation effects on cells, organs, and humans contributes to understanding the side effects of tumor radiotherapy.
- The successful development of nutritional and pharmaceutical countermeasures against radiation risks in human spaceflight such as enriched food or organic thiophosphates fertilize respective efforts in context with cancer therapy on Earth.

Microbiology

- Development of improved technologies for monitoring and identifying microbial contamination in space supports corresponding efforts on Earth, thereby stimulating research on microbial communities and their ecology in natural environments as well as in human-constructed infrastructure such as buildings, ships, cars, trains, and planes.
- The ISS microbiome poses a potential threat for material integrity (biofilm formation) but has not yet developed characteristics that are critical for human health.
- Development of antimicrobial surfaces such as AgXX® with promising applications in space and on Earth.

Protein Crystallization

- Significant progress in structure elucidation of organic macromolecules crystallized in microgravity, such as photosystem I, certain RNA species, and mistletoe lectins with application potential in the treatment of tumors and viral infections as well as for the design of new drugs.
- First successful crystallization of S-layer surface proteins of Archaebacteria with peculiar properties and application potential in biotechnology.
- Significant progress in the understanding of fundamental steps of the crystallization process of macromolecules.

ICARUS Project—Animal Research from Space

- ICARUS was developed to monitor local and global migration of animals from the ISS opening a completely new field of space life sciences research.
- In the future, wide-spread application potential such as in ecology and climate change, for conservation and protection of natural resources up to localizing cargo and international trafficking.

3.6.2   Human Physiology and Space Medicine

Cardiovascular System

- In contradiction to textbook theory, central venous pressure decreases inflight.
- Intraocular pressure is shown to increase rapidly in microgravity then slowly returning to normal level as measured by self-tonometry.
- Impressive cardiovascular adaptation capability to the variable conditions of spaceflight demonstrated—when sufficient training and balanced nutrition is provided, measured by noninvasive portable equipment.
- Investigation of aortic blood pressure and microcirculation of increasing importance with application potential also in the clinical environment for patients with circulatory problems.
- Critically elevated core body temperature and changes in circadian rhythm demonstrated in astronauts—important findings in context with global warming on Earth and for shift workers.
- Textbook knowledge of pulmonary ventilation questioned with significance for therapy management, e.g., during artificial respiration in the clinic.

Vestibular System, Balance Control, and Motor Coordination in Space

- Caloric nystagmus demonstrated to occur also in microgravity disproving the theory of Nobel prize laureate Barany.
- New views on the functioning of the vestibular system of the inner ear with consequences for the role of otolith asymmetry in space motion sickness.
- Endocannabinoids were found to play role in motion sickness, leading possibly to a new therapeutic approach.
- Balance training under partial unloading ("Mars gravity")—established as a new therapy option for elderly people and patients with motor impairments.

Bone and Muscle System

- In general, space missions as time-lapse for degradation in muscles and bones as experienced during the aging process on Earth.
- Significant progress in the elucidation of the mechanisms of bone loss (osteoporosis) and muscle atrophy in microgravity with terrestrial application potential.

- Movement behavior, not spine elongation identified as a cause for back pain in astronauts and subjects of bed rest studies leading to recommendations of preventive strategies in space and on Earth.
- Animal studies in space help elucidate signaling pathways in muscle atrophy, possibly leading to new countermeasure targets.
- First investigations in astronauts and in subjects of bed rest studies confirm a degradation of cartilage provoked by microgravity and immobility, respectively, stimulating the development of effective countermeasures.
- In general, developments of new effective countermeasures promise great application potential for maintaining health and performance as well as in rehabilitation and clinical treatment especially in the aging society of the industrialized countries.

Immune Defense Issues in Space and in Intensive Care on Earth

- No negative impact of spaceflight conditions on the skin of astronauts, if sufficient hygiene is followed.
- Significant results on the functioning of the immune system, especially in elucidating the mechanisms of changes in the innate and adaptive immune system on the cellular level.
- New approaches for health monitoring such as following the immune system status or illness-related changes via breath gas analysis; in addition, online monitoring of the immune status via a newly established cytokine release assay.
- "Inflammaging"—a low-grade, pre-inflammatory state—provoked by stress found in astronauts and in elderly patients thus calling for a joint mitigation strategy.
- Cancer research in space helps to identify new approaches for treatment strategies as well as novel targets for the development of drugs.
- Tissue engineering in microgravity leading to 3D aggregates supports new approaches in regenerative medicine on Earth.

Integrative Physiology in Space Establishing New Concepts for Hypertension Treatment and the Role of Salt

- Dogma of salt storage and the close correlation of sodium with water balance proved to be wrong by space experiments and accompanying ground studies.
- Human skin is established to be an important storage tissue and regulator for sodium.
- Sodium storage involves macrophages which in turn salt-dependently co-regulate blood pressure.
- Sodium showed to strongly influence bone and protein metabolism as well as immune functions.
- During the aging process body sodium storage is increased—in men more than in women—which in turn might enhance the aging process of the human body.
- Altogether, the results lead to new approaches for counteracting diseases such as hypertension and osteoporosis.

- ^{23}Na magnetic resonance spectroscopy was established as new noninvasive method to visualize sodium in human tissues supporting the diagnosis of diseases such as sclerosis, hypertension, and diabetes.

Human Health and Performance Issues

- Exercise found to benefit not only the physiology of humans but also mental health in space and on Earth.
- Individualization of exercise programs with considering personal preferences demonstrated to have important impacts on training success.
- Standardized psychophysiological measurements of stress effects and wireless monitoring of crew relations enabled by new noninvasive devices ("Healthlab").
- Sleep monitoring via simple assessment of the autonomic nervous system by "Somnowatch," applicable also in extreme environments and in occupational health.

Space Life Sciences Research and Aging on Earth

- In general, the physiological and psychological alterations and challenges experienced by astronauts in spaceflight or by subjects in bed rest studies demonstrated to resemble those of the aging process on Earth.
- Since the changes in space occur in time-lapse, research results as well as testing of countermeasures achievable much faster for the benefit of people on Earth.
- In total, spaceflight and bed rest studies as extremes of modern sedentary lifestyles and models for aging providing a new understanding of the respective health consequences—thereby hopefully leading to a unified approach in integrative physiology in space and in terrestrial medicine to finding solutions for maintaining health and mobility, as we live longer: "Geriatrics meets Spaceflight."

References

Abeln V, MacDonald-Nethercott E, Piacentini MF, Meeusen R, Kleinert J, Strueder HK, Schneider S (2015) Exercise in isolation—a countermeasure for electrocortical, mental and cognitive impairments. PLoS One. https://doi.org/10.1371/journalpone0126356

Abeln V, Vogt T, Schneider S (2016) Neurocognitive and neuro-affective effects of exercise. In: Schneider S (ed) Exercise in space, SpringerBriefs in space life sciences. Springer, Cham. https://doi.org/10.1007/978-3-319-29571-8_5

Ahmed HAM, Häder DP (2011) Monitoring of waste water samples using the ECOTOX biosystem and the flagellate alga Euglena gracilis. Chemistry Water Air Soil Pollut. https://doi.org/10.1007/S11270-010-0552-4

Andreev-Andrievskiy A, Popova A, Boyle R, Alberts J, Shenkman B, Vinogradova O, Dolgov O, Anokhin K, Tsvirkun D, Soldatov P, Nemirovskaya T, Ilyin E, Sychev V (2014) Mice in Bion-M 1 Space mission: training and selection. PLoS One 9:1–15. https://doi.org/10.1371/journal.pone.0104830

Baevsky RM, Funtova IL, Diedrich A, Chernikova AG, Drescher J, Baranov VM, Tank J (2009) Autonomic function testing aboard the ISS using "PNEUMOCARD". Acta Astronaut 65:930–932

Baevsky RM, Funtova II, Luchitskaya ES, Chernikova AG (2014) The effects of long-term microgravity on autonomic regulation of blood circulation in crewmembers of the international space station. Cardiometry 5:35–49

Bailey JF, Miller SL, Khieu K, O'Neill CWO, Healey RM, Coughlin DG, Sayson JV, Chang DG, Hargens AR, Lotz JC (2018) From the international space station to the clinic: how prolonged unloading may disrupt lumbar spine stability. Spine J 18:7–14. https://doi.org/10.1016/j.spinee. 2017.08.261

Barger LK, Dinges DF, Czeisler CA (2020) Sleep and circadian effects of space. In: Young LR, Sutton JP (eds) Handbook of bioastronautics. Springer, Cham. https://doi.org/10.1007/978-3-319-10152-1_86-2

Basner M, Rao H, Goel N, Dinges DF (2013) Sleep deprivation and neurobehavioral dynamics. Curr Opin Neurobiol 23:854–863

Baum K, Essfeld D (1999) Origin of back pain during bedrest: a new hypothesis. Eur J Med Res 4:389–393

Baum K, Hoy S, Essfeld D (1997) Continuous monitoring of spine geometry: a new approach to study back pain in space. Int J Sports Med 18:331–333

Becker JL, Souza GR (2013) Using space-based investigations to inform cancer research on Earth. Nat Rev Cancer 13:315–327

Belavy D, Adams M, Brisby H, Cagnie B, Danneels L, Fairbank J, Hargens AR, Judex S, Scheuring RA, Sovelius R, Urban J, van Dieën JH, Wilke HJ (2016) Disc herniations in astronauts: what causes them, and what does it tell us about herniation on Earth? Eur Spine J 25:144–154. https://doi.org/10.1007/s00586-015-3917-y

Berger T, Bilski P, Hajek M, Puchalska M, Reitz G (2013) The MATROSHKA experiment: results and comparison from extravehicular activity (MTR-1) and intravehicular activity (MTR-2A/2B) exposure. Radiat Res 180:622–637

Bimpong-Buta NY, Jirak P, Wernly B, Lichtenauer M, Knost T, Abusamrah T, Kelm M, Jung C (2018) Blood parameter analysis after short term exposure to weightlessness in parabolic flight. Clin Hemorheol Microcirc 70:477–486. https://doi.org/10.3233/CH-189314

Bimpong-Buta NY, Muessig JM, Knost T, Masyuk M, Binneboessel S, Nia AM, Kelm M, Jung C (2020) Comprehensive analysis of macrocirculation and microcirculation in microgravity during parabolic flights. Front Physiol 11:960. https://doi.org/10.3389/fphys.2020.00960

Bloomfield SA (2020) Bone loss. In: Young LR, Sutton JP (eds) Handbook of bioastronautics. Springer, Cham. https://doi.org/10.1007/978-3-319-10152-1_95-2

Blottner D, Salanova M (2015) The neuromuscular system: from earth to space life sciences, SpringerBriefs in space life sciences. Springer, Cham

Blüm V (2003) Aquatic modules for bioregenerative life support systems: developmental aspects based on the space flight results of the C.E.B.A.S. mini-module. Adv Space Res 48:792–798

Bogaerts A, Delecluse C, Claessens AL, Coudyzer W, Boonen S, Verschueren SMP (2007) Impact of whole-body vibration training versus fitness training on muscle strength and muscle mass in older men: a 1-year randomized controlled trial. J Gerontol A Biol Sci Med Sci 62:630–635. https://doi.org/10.1093/gerona/62.6.630

Boyle R, Hughes-Fulford M (2020) Space biology (cells to amphibians). In: Young LR, Sutton JP (eds) Handbook of bioastronautics. Springer, Cham. https://doi.org/10.1007/978-3-319-10152-1_39-2

Braun M, Limbach C (2006) Rhizoids and protonemata of characean algae: model cells for research on polarized growth and plant gravity sensing. Protoplasma 229:133–142

Braun M, Böhmer M, Häder DP, Hemmersbach R, Palme K (2018) Gravitational biology I: Gravity sensing and graviorientation in microorganisms and plants, SpringerBriefs in space life sciences. Springer, Cham. https://doi.org/10.1007/978-3-319-93894-3

Braun N, Binder S, Grosch H, Theek C, Ülker J, Tronnier H, Heinrich U (2019a) Current data on effects of long-term missions on the International Space Station on skin physiological parameters. Skin Pharmacol Physiol 32:43–51

Braun N, Thomas S, Tronnier H, Heinrich U (2019b) Self-reported skin changes by a selected number of astronauts after long-duration mission on ISS as part of the Skin B project. Skin Pharmacol Physiol 32:52–57. https://doi.org/10.1159/000494689

Brümmer V, Schneider S, Abel T, Vogt T, Strüder HK (2011) Brain cortical activity is influenced by exercise mode and intensity. Med Sci Sports Exerc 43:1863–1872

Buchheim JI, Hoskyns S, Moser D, Han B, Deindl E, Hörl M, Moser D, Biere K, Feuerecker M, Schelling G, Thieme D, Kaufmann I, Thiel M, Choukèr A (2018) Oxidative burst and dectin-1-triggered phagocytosis affected by norepinephrine and endocannabinoids: implications for fungal clearance under stress. Int Immunol 30:79–89

Buchheim JI, Matzel S, Rykova M, Vassilieva G, Ponomarev S, Nichiporuk I, Hörl M, Moser D, Biere K, Feuerecker M, Schelling G, Thieme D, Kaufmann I, Thiel M, Choukèr A (2019) Stress related shift toward inflammaging in cosmonauts after long-duration space flight. Front Physiol. https://doi.org/10.3389/fphys.2019.00085

Buchheim JI, Crucian B, Choukèr A (2020) Immunology. In: Young LR, Sutton JP (eds) Handbook of bioastronautics. Springer, Cham. https://doi.org/10.1007/978-3-319-10152-1_23-2

Caiozzo VJ, Baldwin KM (2020) Muscle wasting in space and countermeasures. In: Young LR, Sutton JP (eds) Handbook of bioastronautics. Springer, Cham. https://doi.org/10.1007/978-3-319-10152-1_70-1

Cappuccio FP (2013) Cardiovascular and other effects of salt consumption. Kidney Int Suppl 3 (4):312–315. https://doi.org/10.1038/kisup.2013.65

Cheron G, Leroy A, De Saedeleer C, Bengoetxea A, Lipshits M, Cebolla A, Servais L, Dan B, Berthoz A, McIntyre J (2006) Effect of gravity on human spontaneous 10-Hz electroencephalographic oscillations during the arrest reaction. Brain Res 1121:104–116. https://doi.org/10.1016/j.brainres.2006.08.098

Choukèr A (2020) Stress challenges and immunity in space – from mechanisms to monitoring and preventive strategies, 2nd edn. Springer, Berlin

Choukèr A, Stahn AC (2020) COVID-19 - the largest isolation study in history: the value of shared learnings from spaceflight analogs. NPJ Microgravity 6:32. https://doi.org/10.1038/s41526-020-00122-8

Choukèr A, Ullrich O (2016) The immune system in space: are we prepared? SpringerBriefs in space life sciences. Springer, Cham. https://doi.org/10.1007/978-3-319-41466-9

Choukèr A, Kaufmann I, Kreth S, Hauer D, Feuercker M, Thieme D, Vogeser M, Thiel M, Schelling G (2010) Motion sickness, stress and the endocannabinoid system. PLoS One 5: e10752

Clarke AH (2008) Listing's plane and the otolith-mediated gravity vector. Progr Brain Res 171:291–294

Clarke AH (2017) Vestibulo-oculomotor research in space, SpringerBriefs in space life sciences. Springer, Cham. https://doi.org/10.1007/978-3-319-59933-5

Clarke AH, Schönfeld U (2015) Modification of unilateral otolith responses following spaceflights. Exp Brain Res 233:3613–3624

Clarke AH, Just K, Krzok W, Schönfeld U (2013) Listing's plane and the 3D-VOR in microgravity – the role of the otolith afferences. J Vestib Res 23:61–70

Claus H, Akca E, Debaerdemaeker T, Evrard C, Declercq JP, König H (2002) Primary structure of selected archaeal mesophilic and extremely thermophilic outer surface layer proteins. Syst Appl Microbiol 25:3–12. https://doi.org/10.1078/0723-2020-00100

Clément G, Ngo-Anh JT (2013) Space physiology II: Adaptation of the central nervous system to space flight - past, current, and future studies. Eur J Appl Physiol 113:1655–1672

Cochrane DJ (2011) Vibration exercise: the potential benefits. Int J Sports Med 32:75–99

Convertino VA, Ryan KL, Rickards CA, Glorsky SL, Idris AH, Yannopoulos D, Metzger A, Lurie
    KG (2011) Optimizing the respiratory pump: harnessing inspiratory resistance to treat systemic
    hypotension. Respir Care 56:846–857
Cottin H, Rettberg P (2019) EXPOSE-R2 on the International Space Station. Astrobiology 19(8).
    https://doi.org/10.1089/ast.2019.0625
Cottin H, Kotler JM, Bartik K, Cleaves HJ, Cockell CS, de Vera JP, Ehrenfreund P, Leuko S, Ten
    Kate IL, Martins Z, Pascal R, Quinn R, Rettberg P, Westall F (2015) Astrobiology and the
    possibility of life on Earth and elsewhere. Space Sci Rev 209:1–42
Cottin H, Kotler JM, Billi D, Cockell C, Demets R, Ehrenfreund P, Elsaesser A, d'Hendecourt K,
    van Loon JWA, Martins Z, Onofri S, Quinn RC, Rabbow E, Rettberg P, Ricco AJ, Slenzka K,
    de la Torre R, de Vera JP, Westall F, Carrasco N, Fresneau A, Kawaguchi Y, Kebukawa Y,
    Nguyen D, Poch O, Saiagh K, Stalport F, Yamagishi A, Yano A, Klamm BA (2017) Space as a
    tool for astrobiology: review and recommendations for experimentations in Earth orbit and
    beyond. Space Sci Rev 209:83–181. https://doi.org/10.1007/s11214-017-0365-5
Crandall CG, Johnson JM, Convertino VA, Raven PB, Engelke KA (1994) Altered thermoregula-
    tory responses after 15 days of head-down tilt. J Appl Physiol 77:1863–1867. https://doi.org/10.
    1152/jappl.1994.77.4.1863
Crucian BE, Choukèr A, Simpson RJ, Mehta S, Marshall G, Smith SM, Zwart SR, Heer M,
    Ponomarev S, Whitmire A, Frippiat JP, Douglas GL, Lorenzi H, Buchheim JI, Makedonas G,
    Ginsburg GS, Ott CM, Pierson DL, Krieger SS, Baecker N, Sams C (2018) Immune system
    dysregulation during spaceflight: potential countermeasures for deep space exploration mis-
    sions. Front Immunol. https://doi.org/10.3389/fimmu.2018.01437
De Vera JP (2020) Astrobiology on the International Space Station, SpringerBriefs in space life
    sciences. Springer, Cham. https://doi.org/10.1007/978-3-030-61691-5
Demêmes D, Dechesne CJ, Ventéo S, Gaven F, Raymond J (2001) Development of the rat efferent
    vestibular system on the ground and in microgravity. Dev Brain Res 128:35–44. https://doi.org/
    10.1016/s0165-3806(01)00146-8
Demontis GC, Germani MM, Caiani EG, Barravecchia I, Passino C, Angeloni D (2017) Human
    pathophysiological adaptations to the space environment. Front Physiol. https://doi.org/10.
    3389/fphys.2017.00547
Dick R, Penzel T, Fietze I, Partinen M, Hein H, Schulz J (2010) AASM standards of practice
    compliant validation of actigraphy sleep analysis from SOMNOwatch™ versus
    polysomnographic sleep diagnostics shows high conformity also among subjects with sleep
    disordered breathing. Physiol Meas 31:1623–1633
Dijk DJ, Neri DF, Wyatt JK, Ronda JM, Riel E, Ritz-De Cecco A, Hughes RJ, Elliott AR, Prisk GK,
    West JB, Czeisler CA (2001) Sleep, performance, circadian rhythms, and light-dark cycles
    during two space shuttle flights. Am J Physiol Regul Integr Comp Physiol 281:R1647–R1664.
    https://doi.org/10.1152/ajpregu.2001.281.5.R1647
Dolch ME, Hummel T, Fetter V, Helwig A, Schelling G (2017) Electronic nose, functionality for
    breath gas analysis during parabolic flight. Microgr Sci Technol 29:201–207
Draeger J, Schwartz R, Groenhoff S, Stern C (1993) Self-tonometry under microgravity conditions.
    Clin Invest 71:700–703
Draeger J, Schwartz R, Groenhoff S, Stern C (1994) Self-tonometry during the German 1993
    Spacelab D-2 mission. Ophthalmology 91:697–699
Dreiner M, Willwacher S, Kramer A, Kümmel J, Frett T, Zaucke F, Liphardt AM, Gruber M,
    Niehoff A (2020) Short-term response of serum cartilage oligomeric matrix protein to different
    types of impact loading under normal and artificial gravity. Front Physiol 11:1032. https://doi.
    org/10.3389/fphys.2020.01032
Drummer C, Hesse C, Baisch F, Norsk P, Elmann-Larsen B, Gerzer R, Heer M (2000) Water and
    sodium balances and their relation to body mass changes in microgravity. Eur J Clin Investig
    30:1066–1075

Edgerton VR, Roy RR, Recktenwald MR, Hodgson JA, Grindeland RE, Kozlovskaya I (2000) Neural and neuroendocrine adaptations to microgravity and ground-based models of microgravity. J Gravit Physiol 7:45–52

Ekkekakis P (2009) Let them roam free? Physiological and psychological evidence for the potential of self-selected exercise intensity in public health. Sports Med 39:857–888

Ekkekakis P, Acevedo EO (2006) Affective responses to acute exercise: toward a psychobiological dose-response model. In: Acevedo EO, Ekkekakis P (eds) Psychobiology of physical activity. Human Kinetics, Champaign, IL, pp 91–109

Evrard C, Declercq J-P, Debaerdemaeker T, König H (1999) The first successful crystallization of a prokaryotic extremely thermophilic outer surface layer glycoprotein. Z Krist 214:427–442

Feuerecker M, Mayer W, Kaufmann I, Gruber M, Muckenthaler F, Yi B, Salam AP, Briegel J, Schelling G, Thiel M, Choukèr A (2012) A corticoid-sensitive cytokine release assay for monitoring stress-mediated immune modulation. Clin Exp Immunol. https://doi.org/10.1111/cei.12049

Feuerecker M, Sudhoff L, Crucian B, Pagel JI, Sams C, Strewe C, Guo A, Schelling G, Briegel J, Kaufmann I, Choukèr A (2018) Early immune anergy towards recall antigens and mitogens in patients at onset of septic shock. Sci Rep. https://doi.org/10.1038/s41598-018-19976-w

Feuerecker M, Mayer W, Gruber M, Muckenthaler F, Draenert R, Bogner J, Kaufmann I, Crucian B, Rykova M, Morukov B, Sams, C, Choukèr A (2019) Particular characterization of an in-vitro-DTH test to monitor cellular immunity – applications for patient care and space flight. https://ntrs.nasa.gov/citations/20130011428

Fietze I, Penzel T, Partinen M, Sauter J, Küchler G, Suvoro A, Hein H (2015) Actigraphy combined with EEG compared to polysomnography in sleep apnea patients. Physiol Meas 36:385

Foldager N, Andersen TA, Jessen FB, Ellegaard P, Stadeager C, Videbaek R, Norsk P (1996) Central venous pressure in humans during microgravity. J Appl Physiol 81:408–412

Fowler B, Manzey D (2000) Summary of research issues in monitoring of mental health and perceptual-motor performance and stress in space. Aviation Space Environ Med 71:A76–A77

Frerichs I, Dudykevych T, Hinz J, Bodenstein M, Hahn G, Hellige G (2001) Gravity effects on regional lung ventilation determined by functional EIT during parabolic flights. J Appl Physiol 91:39–50

Frerichs I, Pulletz S, Elke G, Zick G, Weiler N (2010) Electrical impedance tomography in acute respiratory distress syndrome. Open Nucl Med J 2:110–118

Freyler K, Weltin E, Gollhofer A, Ritzmann R (2014) Improved postural control in response to a 4-week balance training with partially unloaded bodyweight. Gait Posture 40:291–296

Frings-Meuthen P, Baecker N, Heer M (2008) Low-grade metabolic acidosis may be the cause of sodium chloride-induced exaggerated bone resorption. J Bone Miner Res 23:517–524

Frings-Meuthen P, Luchitskaya E, Jordan J, Tank J, Lichtinghagen R, Smith SC, Heer M (2020) Natriuretic peptide resetting in astronauts. Circulation 141:1593–1595

Garrett-Bakelman FE, Darshi M, Green SJ, Gur RC, Lin L, Macias BR, McKenna MJ, Meydan C, Mishra T, Nasrini J, Piening BD, Rizzardi LF, Sharma K, Siamwala JH, Taylor L, Vitaterna MH, Afkarian M, Afshinnekoo E, Ahadi S, Ambati A, Arya M, Bezdan D, Callahan CM, Chen S, Choi AMK, Chlipala GE, Contrepois K, Covington M, Crucian BE, De Vivo I, Dinges DF, Ebert DJ, Feinberg JI, Gandara JA, George KA, Goutsias J, Grills GS, Hargens AR, Heer M, Hillary RP, Hoofnagle AN, Hook VYH, Jenkinson G, Jiang P, Keshavarzian A, Laurie SS, Lee-McMullen B, Lumpkins SB, MacKay M, Maienschein-Cline MG, Melnick AM, Moore TM, Nakahira K, Patel HH, Pietrzyk R, Rao V, Saito R, Salins DN, Schilling JM, Sears DD, Sheridan CK, Stenger MB, Tryggvadottir R, Urban AE, Vaisar T, Van Espen B, Zhang J, Ziegler MG, Zwart SR, Charles JB, Kundrot CE, Scott GBI, Bailey SM, Basner M, Feinberg AP, Lee SMC, Mason CE, Mignot E, Rana BK, Smith SM, Snyder MP, Turek FW (2019) The NASA Twins Study: a multidimensional analysis of a year-long human spaceflight. Science. https://doi.org/10.1126/science.aau8650

Gerzer (2014) Salt balance: from space experiments to revolutionizing new clinical concepts on Earth – a historical review. Acta Astronaut 104:378–382

Goswami N (2017) Falls and fall prevention in older persons: geriatrics meets spaceflight. Front Physiol 8:603. https://doi.org/10.3389/fphys.2017.00603

Goswami N, White O, van Loon JJWA, Roessler A, Blaber AP (eds) (2020) Gravitational physiology, aging and medicine. Front Physiol. https://doi.org/10.3389/978-2-88963-273-2

Grimm D, Wehland M, Pietsch J, Aleshcheva G, Wise P, van Loon J, Ulbrich C, Magnusson NE, Infanger M, Bauer J (2017) Growing tissues in real and simulated microgravity: new methods for tissue engineering. Tissue Eng Part B 20:555–566. https://doi.org/10.1089/ten.TEB.2013.070

Groenhoff S, Draeger J, Deutsch C, Wiezorrek R, Hock B (1992) Self-tonometry: technical aspects of calibration and clinical application. Int Ophthalmol 16:299–303

Gundel A, Polyakov VV, Zulley J (1997) The alteration of human sleep and circadian rhythms during spaceflight. J Sleep Res 6:1–8

Gunga HC (2020) Am Tag zu heiss und nachts zu hell: Was unser Körper kann – und warum er heute überfordert ist. Rowohlt, Hamburg. ISBN 978-3-498-02540-3

Gunga HC, Weller von Ahlefeld V, Appell Coriolano HJ, Werner A, Hoffmann U (2016) Cardiovascular system, red blood cells, and oxgen transport in microgravity, SpringerBriefs in space life sciences. Springer, Cham. https://doi.org/10.1007/978-3-319-33226-0

Guo SS, Mao RX, Zhang LL, Tang YK, Li YH (2017) Progress and prospect of research on controlled ecological life support technique. REACH Rev Hum Space Explor 6:1–10

Häder DP, Erzinger GS (2015) Advanced methods in image analysis as potent tools in online biomonitoring of water resources. Recent Top Pat Imaging 5:112–118

Häder DP, Hemmersbach R, Lebert M (2005) Gravity and the behavior of unicellular organisms. Cambridge University Press, Cambridge

Häder DP, Braun M, Hemmersbach R (2018) Bioregenerative life support systems in space research. Gravitational biology I, SpringerBriefs in space life sciences. Springer, Cham. https://doi.org/10.1007/978-3-319-93894-3_8

Hanke WQ, Kohn FPM, Neef M, Hampp R (2018) Gravitational Biology II. Interaction of gravity with cellular components and cell metabolism, SpringerBriefs in space life sciences. Springer, Cham. https://doi.org/10.1007/978-3-030-00596-2

Hauschild S, Tauber S, Lauber B, Thiel CS, Layer LE, Ullrich O (2014) T cell regulation in microgravity – the current knowledge from in vitro experiments conducted in space, parabolic flights and ground-based facilities. Acta Astronaut 104:365–377

Hauschild S, Tauber S, Lauber BA, Thiel CS, Layer LE, Ullrich O (2016) Cellular effects of altered gravity on the human adaptive immune system. In: The immune system in space: are we prepared? SpringerBriefs in space life sciences. Springer, Cham. https://doi.org/10.1007/978-3-319-41466-9_5

Heer M, Baisch F, Kropp J, Gerzer R, Drummer C (2000) High dietary sodium chloride consumption may not induce body fluid retention in humans. Am J Phys 278:F585–F595

Heer M, Titze J, Smith SM, Baecker N (2015) Nutrition, physiology and metabolism in spaceflight and analog studies, SpringerBriefs in space life sciences. Springer, Cham

Hellweg CE, Berger T, Matthiä D, Baumstark-Khan C (2020) Radiation in space: relevance and risk for human missions, SpringerBriefs in space life sciences. Springer, Cham. https://doi.org/10.1007/978-3-030-46744-9

Hilbig RW, Anken RH (2017) Impact of micro- and hypergravity on neurovestibular issues in fish. In: Sensory motor and behavioral research in space, SpringerBriefs in space life sciences. Springer, Cham. https://doi.org/10.1007/978-3-319-68201-3_4

Holubarsch J, Helm M, Ringhof S, Gollhofer A, Freyler K, Ritzmann R (2019) Tumbling reactions in hypo and hyper gravity - muscle synergies are robust across different perturbations of human stance during parabolic flights. Sci Rep 9:10490. https://doi.org/10.1038/s41598-019-47091-x

Horneck G, Zell M (2012) The EXPOSE-E mission. Astrobiology 12(5). https://doi.org/10.1089/ast.2012.0831

Horneck G, Panitz C, Zell M (2015) Expose-R. Int J Astrobiol 14(1). https://doi.org/10.1017/51473550414000xxx

Horstmann M, Durante M, Johannes C, Pieper R, Obe G (2005) Space radiation does not induce a significant increase of intrachromosomal exchange in astronauts' lymphocytes. Radiat Environ Biophys 44:219–224

Ichijo T, Yamaguchi N, Tanigaki F, Shirakawa M, Nasu M (2016) Four-year bacterial monitoring in the international space station Japanese experiment module kibo with culture-independent approach. NPJ Microgravity 2. https://doi.org/10.1038/npjmgrav.2016.7

Ingber DE (2008) Tensegrity-based mechanosensing from macro to micro. Progr Biophys Mol Biol 97:163–179

Jamon M (2014) The development of vestibular system and related functions in mammals: impact of gravity. Front Integr Neurosci 8(11):1–13. https://doi.org/10.3389/fnint.2014.00011

Johannes B, Salnitski VP, Polyakov VV, Kirsch KA (2003) Changes in the autonomic reactivity pattern to psychological load under long-term microgravity—twelve men during 6-month spaceflights. Aviakosm Ekolog Med 37:6–16

Johannes B, Salnitski V, Soll H, Rauch M, Hoermann HJ (2008) De-individualized psychophysiological strain assessment during a flight simulation test—validation of a space methodology. Acta Astronaut 63:791–799

Johannes B, Sitev AS, Vinokhodova AG, Salnitski VP, Savchenko EG, Artyukhova AE, Bubeev YA, Morukov BV, Tafforin C, Basner M, Dinges DF, Rittweger J (2015) Wireless monitoring of changes in crew relations during long-duration mission simulation. PLoS One. https://doi.org/10.1371/journal.pone.0134814

Johannes B, Salnitski V, Dudukin A, Shevchenko L, Bronnikov S (2016a) Performance assessment in the PILOT experiment on board space station MIR and ISS. Aerospace Med Hum Perf 87:534–544

Johannes B, Bernius P, Lindemann J, Kraus de Camargo O, Oerter R (2016b) Feasibility study using in-water EEG measurement during dolphin assisted therapy (DAT). Int J Clin Psychiatry 4:17–25

Johnson CM, Subramanian A, Pattathil S, Correll MJ, Kiss JZ (2017) Comparative transcriptomics indicate changes in cell wall organization and stress response in seedlings during spaceflight. Am J Bot 104(8):1219–1231

Johnston SL, Campbell MR, Scheuring R, Feiveson AH (2010) Risk of herniated nucleus pulposus among U.S. astronauts. Aviat Space Environ Med 81:566–574

Kanas N (2020) Crewmember interactions in space. In: Young LR, Sutton JP (eds) Handbook of bioastronautics. Springer, Cham. https://doi.org/10.1007/978-3-319-10152-1_33-2

Kaufmann I, Draenert R, Gruber M, Feuerecker M, Roider J, Choukèr A (2013) A new cytokine release assay: a simple approach to monitor the immune status of HIV-infected patients. Infection. https://doi.org/10.1007/s15010-013-0445-8

Kern P (2009) Biological life support systems. In: Ley W, Wittmann K, Hallmann W (eds) Handbook of space technology. Wiley, Hoboken, NJ

Kirsch KA, Röcker L, Gauer OH, Krause R (1984) Venous pressure in man during weightlessness. Science 225:218–219

Kirsch KA, Haenel F, Röcker L (1986) Venous pressure in microgravity. Naturwissenschaften 73:447–449

Kiss JZ, Wolverton C, Wyatt SE, Hasenstein KH, Loon JJWA (2019) Comparison of microgravity analogs to spaceflight in studies of plant growth and development. Front Plant Sci 10:1577. https://doi.org/10.3389/fpls.2019.01577

Kitada K, Daub S, Zhang Y, Klein JD, Nakano D, Pedchenko T, Lantier L, LaRocque L, Marton A, Neubert P, Schröder A, Rakova N, Jantsch J, Dikalova AE, Dikalov SI, Harrison DC, Müller DN, Nishiyama A, Rauh M, Harris RC, Luft FC, Wasserman DH, Sands J, Titze J (2017) High salt intake reprioritizes osmolyte and energy metabolism for body fluid conservation. J Clin Invest 127:1944–1959

Klein T, Wollseiffen P, Sanders M, Claassen J, Carnahan H, Abeln V, Vogt T, Strüder HK, Schneider S (2019) The influence of microgravity on cerebral blood flow and electrocortical activity. Exp Brain Res 237:1057–1062. https://doi.org/10.1007/s00221-019-05490-6

Klein T, Sanders M, Wollseiffen P, Carnahan H, Abeln V, Askew CD, Claassen JA, Schneider S (2020) Transient cerebral blood flow responses during microgravity. Life Sci Space Res 25:66–71. https://doi.org/10.1016/j.lssr.2020.03.003

Klukas O, Schubert W-D, Jordan P, Krauß N, Fromme P, Witt HT, Saenger W (1999) Photosystem I, an improved model of the stromal subunits PsaC, PsaD, and PsaE. J Biol Chem 274:7351–7360

Konda NN, Karri RS, Winnard A, Nasser M, Evetts S, Boudreau E, Caplan N, Gradwell D, Velho RM (2019) A comparison of exercise interventions from bed rest studies for the prevention of musculoskeletal loss. NPJ Microgravity 5:12. https://doi.org/10.1038/s41526-019-0073-4

Kopp C, Linz P, Wachsmuth L, Dahlmann A, Horbach T, Schöfl C, Renz W, Santoro D, Niendorf T, Müller DN, Neininger M, Cavallaro A, Eckardt KU, Schmieder RE, Luft FC, Uder M, Titze J (2012) Na magnetic resonance imaging of tissue sodium. Hypertension 59:167–172

Kopp C, Linz P, Dahlmann A, Hammon M, Jantsch J, Müller DN, Schmieder RE, Cavallaro A, Eckardt KU, Uder M, Luft FC, Titze J (2013) ^{23}Na magnetic resonance imaging-determined tissue sodium in healthy subjects and hypertensive patients. Hypertension 61:635–640

Kopp C, Beyer C, Linz P, Dahlmann A, Hammon M, Jantsch J, Neubert P, Rosenhauer D, Müller DN, Cavallaro A, Eckardt KU, Schett G, Luft FC, Uder M, Distler JHW, Titze J (2017) Na^{+} deposition in the fibrotic skin of systemic sclerosis patients detected by ^{23}Na-magnetic resonance imaging. Rheumatology 56:556–560

Kopp C, Linz P, Maier C, Wabel P, Hammon M, Nagel AM, Rosenhauer D, Horn S, Uder M, Luft FC, Titze J, Dahlmann A (2018) Elevated tissue sodium deposition in patients with type 2 diabetes on hemodialyis detected by ^{23}Na magnetic resonance imaging. Kidney Int 93:1191–1197

Krauspenhaar R, Rypniewski W, Kalkura N, Moore K, DeLucas L, Stoeva S, Mikhailov A, Voelter W, Betzel C (2002) Crystallisation under microgravity of mistletoe lectin I from Viscum album with adenine monophosphate and the crystal structure at 1.9 Å resolution. Acta Crystallogr D Biol Crystallogr 58:1704–1707

Krauß N, Schubert W-D, Klukas O, Fromme P, Witt HT, Saenger W (1996) Photosystem I at 4 Å resolution represents the first structural model of a joint photosynthetic reaction centre and core antenna system. Nat Struct Mol Biol 3:965–973

Krüger M, Bauer J, Grimm D (2017) Cancer research in space. In: Ruyters G, Betzel C, Grimm D (eds) Biotechnology in space, SpringerBriefs in space life sciences. Springer, Cham. https://doi.org/10.1007/978-3-319-64054-9_7

Krüger M, Melnik D, Kopp S, Buken C, Sahana J, Bauer J, Wehland M, Hemmersbach R, Corydon TJ, Infanger M, Grimm D (2019) Fighting thyroid cancer with microgravity research. Int J Mol Sci 20:2553. https://doi.org/10.3390/ijms20102553

Lang JM, Coil DA, Neches RY, Brown WE, Cavalier D, Severance M, Hampton-Marcell JT, Gilbert JA, Eisen JA (2017a) A microbial survey of the International Space Station (ISS). PeerJ 5:e4029. https://doi.org/10.7717/peerj.4029

Lang T, van Loon JJWA, Bloomfield S, Vico L, Chopard A, Rittweger J, Kyparos A, Blottner D, Vuori I, Gerzer R, Cavanagh PR (2017b) Towards human exploration of space: the THESEUS review series on muscle and bone research priorities. NPJ Microgravity 3:8. https://doi.org/10.1038/s41526-017-0013-0

Ledderhos C, Rothe S, Gens A, Johannes B (2010) Objective psycho-physiological strain assessment of AWACS crews in simulators and real flight operations. In: Federal Ministry of Defense (ed) Annual military scientific research report 2009, pp 76–77

Lee AG, Mader TH, Gibson CR, Tarver W, Rabiei P, Riascos RF, Galdamez LA, Brunstetter T (2020a) Spaceflight associated neuro-ocular syndrome (SANS) and the neuro-ophthalmologic effects of microgravity: a review and an update. NPJ Microgravity 6. https://doi.org/10.1038/s41526-020-0097-9

Lee S-J, Lehar A, Meir JU, Koch C, Morgan A, Warren LE, Rydzik R, Youngstrom DW, Chandok H, George J, Gogain J, Michaud M, Stoklased TA, Liu Y, Germain-Lee EL (2020b)

Targeting myostatin/activin A protects against skeletal muscle and bone loss during spaceflight. Proc Natl Acad Sci. https://doi.org/10.1073/pnas.2014716117

Liphardt A-M, Mündermann A, Andriacchi TP, Achtzehn S, Heer M, Mester J (2018) Sensitivity of serum concentration of cartilage biomarkers to 21-days of bed rest. J Orthop Res 36:1465–1471. https://doi.org/10.1002/jor.23786

Littke W, John C (1984) Protein single crystal growth under microgravity. Materials 225:203–204. https://doi.org/10.1126/science.225.4658.203

Machnik A, Nehofer W, Jantsch J, Dahlmann A, Tammela T, Machura K, Park JK, Beck FX, Müller DN, Derer W, Goss J, Ziomber A, Dietsch P, Wagner H, van Rooijen N, Kurtz A, Hilgers KF, Alitalo K, Eckardt KU, Luft FC, Kerjaschi D, Titze J (2009) Macrophages regulate salt-dependent volume and blood pressure by a vascular endothelial growth factor-C-dependent buffering system. Nat Med 15:545–552

Manzey D (2000) Monitoring of mental performance during spaceflight. Aviat Space Environ Med 71:A69–A75

Martins Z, Cottin H, Kotler JM, Carrasco N, Cockell CS, de la Torre Noetzel R, Demets R, de Vera JP, d'Hendecourt L, Ehrenfreund P, Elsaesser A, Foing B, Onofri S, Quinn R, Rabbow E, Rettberg P, Ricco AJ, Slenzka K, Stalport F, ten Kate IL, van Loon JJWA, Westall F (2017) Earth as a tool for astrobiology - a European perspective. Space Sci Rev 209:43–81. https://doi.org/10.1007/s11214-017-0369-1

Martirosyan A, DeLucas LJ, Schmidt C, Perbandt M, McCombs D, Cox M, Radka C, Betzel C (2019) Effect of macromolecular mass transport in microgravity protein crystallization. Gravit Space Res 7:1. https://doi.org/10.2478/gsr-2019-0005

Matia I, Gonzales-Camacho F, Herranz R, Kiss JZ, Gasset G, van Loon JJWA, Marco R, Medina FX (2010) Plant cell proliferation and growth are altered by microgravity conditions in spaceflight. J Plant Physiol 167:184–193. https://doi.org/10.1016/j.jplph.2009.08.012

McPherson A, De Lucas LJ (2015) Microgravity protein crystallization. NPJ Microgravity 1:15010. https://doi.org/10.1038/npjmgrav.2015.10

Mermel LA (2013) Infection prevention and control during prolonged human space travel. Clin Infect Dis 56:123–130

Mester J, Kleinöder H, Yue Z (2006) Vibration training: benefits and risks. J Biomech 39:1056–1065

Meyer A, Rypniewski W, Szymański M, Voelter W, Barciszewski J, Betzel C (2008) Structure of mistletoe lectin I from Viscum album in complex with the phytohormone zeatin. Biochim Biophys Acta 1784:1590–1595

Mora M, Wink L, Kögler I, Mahnert A, Rettberg P, Schwendner P, Demets R, Cockell C, Alekhova T, Klingl A, Krause R, Zolotariof A, Alexandrova A, Moissl-Eichinger C (2019) Space station conditions are selective but do not alter microbial characteristics relevant to human health. Nat Commun 10:3990. https://doi.org/10.1038/s41467-019-11682-z

Muthert LWF, Izzo LG, van Zanten M, Aronne G (2020) Root tropisms: investigations on Earth and in space to unravel plant growth direction. Front Plant Sci 10:1807. https://doi.org/10.3389/fpls.2019.01807

Narici MV, de Boer MD (2011) Disuse of the musculo-skeletal system in space and on Earth. Eur J Appl Physiol 111:403–420

Nassef MZ, Kopp S, Weland M, Melnik D, Sahana J, Krüger M, Corydon TJ, Oltmann H, Schmitz B, Schütte A, Bauer TJ, Infanger M, Grimm D (2019) Real microgravity influences the cytoskeleton and focal adhesions in human breast cancer cells. Int J Mol Sci 20:3156. https://doi.org/10.3390/ijms20135156

Niehoff A, Brüggemann GP, Zaucke F, Koo S, Mester J, Liphardt AM (2016) Long-duration space flight and cartilage adaptation: first results on changes in tissue metabolism. Osteoarthr Cartil 24 (Suppl. 1):14–145. https://doi.org/10.1016/j.joca.2016.01.282

Norsk P (2020) Physiological effects of spaceflight – weightlessness: an overview. In: Young LR, Sutton JP (eds) Handbook of bioastronautics. Springer, Cham. https://doi.org/10.1007/978-3-319-10152-1_126-2

Novikova ND (2004) Review of the knowledge of microbial contamination of the Russian manned spacecraft. Microb Ecol 47:127–132

Novikova N, De Boever P, Poddbko S, Deshevaya E, Polikarpov N, Rakova N, Coninx I, Mergeay M (2006) Survey of environmental biocontamination on board the International Space Station. Res Microbiol 157:5–12

Paul AL, Sng NJ, Zupanska AK, Krishnamurthy A, Schultz ER, Ferl RJ (2017) Genetic dissection of the Arabidopsis spaceflight transcriptome: are some responses dispensable for the physiological adaptation of plants to spaceflight? PLoS One. https://doi.org/10.1371/journal.pone.0180186

Paulsen K, Tauber S, Dumrese C, Bradacs G, Simmet DM, Gölz N, Hauschild S, Raig C, Engeli S, Gutewort A, Hürlimann E, Biskup J, Unverdorben F, Rieder G, Hofmänner D, Mutschler L, Krammer S, Buttron I, Philpot C, Huge A, Lier H, Barz I, Engelmann F, Layer LE, Thiel CS, Ullrich O (2015) Regulation of ICAM-1 in cells of the monocyte/macrophage system in microgravity. Biomed Res Int. https://doi.org/10.1155/2015/538786

Pavy Le Traon A, Heer M, Narici MV, Rittweger J (2007) From space to Earth: advances in human physiology from 20 years of bed rest studies (1986–2006). Eur J Appl Physiol 101:143–194

Penzel T, Kantelhardt JW, Bartsch RP, Riedl M, Kraemer JF, Wessel N, Garcia C, Glos M, Fietze I, Schöbel C (2016) Modulations of heart rate, ECG, and cardio-respiratory coupling observed in polysomnography. Front Physiol. https://doi.org/10.3389/fphys.2016.00460

Petit G, Cebolla AM, Fattinger S, Petieau M, Summerer L, Cheron G, Huber R (2019) Local sleep-like events during wakefulness and their relationship to decreased alertness in astronauts on ISS. NPJ Microgravity 10. https://doi.org/10.1038/s41526-019-0069-0

Prasad B, Richter P, Vadakedath N, Mancinelli R, Krüger M, Strauch S, Grimm D, Darriet P, Chapel J-P, Cohen J, Lebert M (2020) Exploration of space to achieve scientific breakthroughs. Biotechnol Adv. https://doi.org/10.1016/j.biotechadv.2020.107572

Preston LJ, Rothschild LJ (2020) Astrobiology: an overview. In: Young LR, Sutton JP (eds) Handbook of bioastronautics. Springer, Cham. https://doi.org/10.1007/978-3-319-10152-1_119-2

Prisk K (2014) Microgravity and the respiratory system. Eur Respir J 43:1459–1471

Rakova N, Jüttner K, Dahlmann A, Schröder A, Linz P, Kopp C, Rauh M, Goller U, Beck L, Agureev A, Vassilieva G, Lenkova L, Johannes B, Wabel P, Moissl U, Vienken J, Gerzer R, Eckardt KU, Müller DN, Kirsch K, Morukov B, Luft FC, Titze J (2013) Long-term space flight simulation reveals infradian rhythmicity in human Na$^+$ balance. Cell Metab 17:125–131

Rakova N, Kitada K, Lerchl K, Dahlmann A, Birukov A, Daub S, Kopp C, Pedchenko T, Zhang Y, Beck L, Johannes B, Marton A, Müller DN, Rauh M, Luft FC, Titze J (2017) Increased salt consumption induces body water conservation and decreases fluid intake. J Clin Invest 127:1932–1943

Rennie MJ, Selby A, Atherton P, Smith K, Kumar V, Glover EL, Philips SM (2010) Facts, noise and wishful thinking: muscle protein turnover in aging and human disuse atrophy. Scand J Med Sci Sports 20:5–9. https://doi.org/10.1111/j.1600-0838.2009.00967.x

Reschke MF, Bloomberg JJ, Harm DL, Paloski WH (1994) Space flight and neurovestibular adaptation. J Clin Pharm 34:609–617

Risin D, Stepaniak PC (2013) Biomedical results of the space shuttle program. NASA SP-2013-607

Ritzmann R, Kramer A, Bernhardt S, Gollhofer A (2014) Whole body vibration training – improving balance control and muscle endurance. PLoS One 26:e89905

Ritzmann R, Freyler K, Weltin E, Krause A, Gollhofer A (2015) Load dependency of postural control – kinematic and neuromuscular changes in response to over and under load conditions. PLoS One 10(6):e0128400. https://doi.org/10.1371/journal.pone.0128400

Ritzmann R, Freyler K, Krause A, Gollhofer A (2016) Bouncing on Mars and the Moon - the role of gravity on neuromuscular control: correlation of muscle activity and rate of force development. J Appl Physiol 121(5):1187–1195. https://doi.org/10.1152/japplphysiol.00692.2016

Ritzmann R, Gollhofer A, Freyler K (2017) Posture and locomotion. In: Hilbig R, Gollhofer A, Bock O, Manzey D (eds) Sensory motor and behavioral research in space, SpringerBriefs in space life sciences. Springer, Cham. https://doi.org/10.3389/fphys.2019.00576

Ritzmann R, Freyler K, Helm M, Holubarsch J, Gollhofer A (2019) Stumbling reactions in partial gravity - neuromechanics of compensatory postural responses and inter-limb coordination during perturbation of human stance. Front Physiol 10:576. https://doi.org/10.3389/fphys.2019.00576

Roberts DR, Stahn AC, Seidler RD, Wuyts FL (2020) Towards understanding the effects of spaceflight on the brain. Lancet Neurol 19(10):808. https://doi.org/10.1016/S1474-4422(20)30304-5

Ronca AE, Moyer EL, Talyansky Y, Lowe M, Padmanabhan S, Choi S, Gong C, Cadena SM, Stodieck L, Globus RK (2019) Behavior of mice aboard the International Space Station. Sci Rep 9:4717. https://doi.org/10.1038/s41598-019-40789-y

Rosnet E, Le Scanff C, Sagal MS (2000) How self-image and personality influence performance in an isolated environment. Environ Behav 32:18–31

Ruyters G, Braun M (2014) Plant biology in space: recent accomplishments and recommendations for future research. Plant Biol 16(Suppl 1):4–11. https://doi.org/10.1111/plb.12127

Ruyters G, Stang S (2016) Space medicine 2025 – a vision. REACH Rev Hum Space Explor 1:55–62

Ruyters G, Betzel C, Grimm D (2017) Biotechnology in space, SpringerBriefs in space life sciences. Springer, Cham. https://doi.org/10.1007/978-3-319-64054-9

Sandal G, Leon G, Palinkas L (2006) Human challenges in polar and space environments. Rev Environ Sci Biotechnol 5:281–296

Sandler H, Vernikos J (1986) Inactivity: physiological effects. Academic, New York

Sandonà D, Desaphy JF, Camerino GM, Bianchini E, Ciciliot S, Danieli-Betto D, Dobrowolny G, Furlan S, Germinario E, Goto K, Gutsmann M, Kawano F, Nakai N, Ohira T, Ohno Y, Picard A, Salanova M, Schiffl G, Blottner D, Musarò A, Ohira Y, Betto R, Conte D, Schiaffino S (2012) Adaptation of mouse skeletal muscle to long-term microgravity in the MDS mission. PLoS One 7(3). https://doi.org/10.1371/journal.pone.0033232

Scherer H, Brandt U, Clarke AH, Merbold U, Parker R (1986) European vestibular experiments on the Spacelab-1 mission. Exp Brain Res 64:255–263

Schneider S, Guardiera S, Abel T, Carnahan H, Strüder H (2009) Artificial gravity results in changes in frontal lobe activity measured by EEG tomography. Brain Res 1285:119–126. https://doi.org/10.1016/j.brainres.2009.06.026

Schneider S, Abeln V, Carnahan H, Kleinert J, Piacentini MF, Meeusen R, Strüder H (2010) Exercise as a countermeasure to psycho-physiological deconditioning during long-term confinement. Behav Brain Res 211:208–214. https://doi.org/10.1016/j.bbr.2010.03.034

Schneider S, Abeln V, Popova J, Fomina E, Jacubowski A, Meeusen R, Strüder H (2013) The influence of exercise on prefrontal cortex activity and cognitive performance during a simulated space flight to Mars (MARS500). Behav Brain Res 236:1–7. https://doi.org/10.1016/j.bbr.2012.08.022

Schubert D (2018) System analysis of plant production in greenhouse modules as an integrated part of planetary habitats. PhD thesis, University Bremen

Seibert FS, Bernhard F, Sterbo U, Vairavanathan S, Bauer F, Rohn B, Pagonas N, Babel N, Jankowski J, Westhoff TH (2018) The effect of microgravity on central aortic blood pressure. Am J Hypertension 31:1183. https://doi.org/10.1093/ajh/hpy119

Shelhamer M, Bloomberg J, LeBlanc A, Prisk GK, Sibonga J, Smith SM, Zwart SR, Norsk P (2020) Selected discoveries from human research in space that are relevant to human health on Earth. NPJ Microgravity 6:5. https://doi.org/10.1038/s41526-020-0095-y

Sitja-Rabert M, Rigau Comas D, Fort Vanmeerhaeghe A, Santoyo Medina C, Roqué i Figuls M, Romero-Rodríguez D, Bonfill Cosp X (2012) Whole-body vibration training for patients with neurodegenerative disease. Cochrane Database Syst Rev 2:CD009097. https://doi.org/10.1002/14651858.CD009097.pub2

Sobisch LY, Rogowski KM, Fuchs J, Schmieder W, Vaishampayan A, Oles P, Novikova N, Grohmann E (2019) Biofilm forming antibiotic resistant gram-positive pathogens isolated from surfaces on the International Space Station. Front Microbiol 19. https://doi.org/10.3389/fmicb.2019.00543

Soderpalm AC, Kroksmark AK, Magnusson P, Karlsson J, Tulinius M, Swolin-Eide D (2013) Whole body vibration therapy in patients with Duchenne muscular dystrophy - a prospective observational study. J Musculoskelet Neuronal Interact 13:13–18

Stahn AC, Werner A, Opatz O, Maggioni MA, Steinach M, von Ahlefeld VW, Moore AD Jr, Crucian BE, Smith SM, Zwart SR, Schlabs T, Mendt S, Trippel T, Koralewski E, Koch J, Chouker A, Reitz G, Shang P, Rocker L, Kirsch KA, Gunga HC (2017) Increased core body temperature in astronauts during long-duration space missions. Sci Rep 7:16180. https://doi.org/10.1038/s41598-017-15560-w

Strangman G, Gur RC, Basner M (2020) Cognitive performance in space. In: Young LR, Sutton JP (eds) Handbook of bioastronautics. Springer, Cham. https://doi.org/10.1007/978-3-319-10152-1_70-1

Su S-H, Gibbs NM, Jancewicz AL, Masson PH (2017) Molecular mechanisms of root gravitropism. Curr Biol 27:R964–R972

Sutton JP (2020) Clinical benefits of bioastronautics. In: Young LR, Sutton JP (eds) Handbook of bioastronautics. Springer, Cham. https://doi.org/10.1007/978-3-319-10152-1_79-2

Tank J, Baevsky RM, Funtova II, Diedrich A, Slepchenkova IN, Jordan J (2011) Orthostatic heart rate responses after prolonged space flights. Clin Auton Res 21:121–124

Tauber S, Ullrich O (2016) Cellular effects of altered gravity on the innate immune system and the endothelial barrier. In: The immune system in space: are we prepared? SpringerBriefs in space life sciences, Springer, Cham. https://doi.org/10.1007/978-3-319-41466-9_5

Thiel CS, de Zélicourt D, Tauber S, Adrian A, Franz M, Simmet DM, Schoppmann K, Hauschild S, Krammer S, Christen M, Bradacs G, Paulsen K, Wolf SA, Braun M, Hatton J, Kurtcuoglu V, Franke S, Tanner S, Christoforetti S, Sick B, Hock B, Ullrich O (2017) Rapid adaptation to microgravity in mammalian macrophage cells. Sci Rep 7:43. https://doi.org/10.1038/s41598-017-00119-6

Thiel CS, Tauber S, Lauber B, Polzer J, Seebacher C, Uhl R, Neelam S, Zhang Y, Levine H, Ullrich O (2019) Rapid morphological and cytoskeletal response to microgravity in human primary macrophages. Int J Mol Sci 20:2402. https://doi.org/10.3390/ijms20102402

Titze J, Bauer K, Schafflhuber M (2005) Internal sodium balance in DOCA-salt rats: a body composition study. Am J Physiol Renal Physiol 289:793–802

Vallaza M, Banumathi S, Perbandt M, Moore K, DeLucas L, Betzel C, Erdmann VA (2002) Crystallization and structure analysis of Thermus flavus 5S rRNA helix B. Acta Crystallogr D Biol Crystallogr 58:1700–17003

Vallaza M, Perbandt M, Klussmann S, Rypniewski W, Einspahr HM, Erdmann VA, Betzel C (2004) First look at RNA in L-configuration. Acta Crystallogr Sect D 60:1–7

Vandenbrink JP, Herranz R, Medina FJ, Edelmann RE, Kiss JZ (2016) A novel blue-light phototropic response is revealed in roots of Arabidopsis thaliana in microgravity. Planta 244:1201–1215

Venkateswaran K, La Duc MT, Horneck G (2014a) Microbial existence in controlled habitats and their resistance to space conditions. Microbes Environ. https://doi.org/10.1264/jsme2.ME14032

Venkateswaran K, Vaishampayan P, Cisneros J, Pierson DL, Rogers SO, Perry J (2014b) International space station environmental microbiome—microbial inventories of ISS filter debris. Appl Microbiol Biotechnol 98:6453–6466. https://doi.org/10.1007/s00253-014-5650-6

Verbanck S, Larsson H, Linnarsson D, Prisk GK, West JB, Paiva M (1997) Pulmonary tissue volume, cardiac output, and diffusing capacity in sustained microgravity. J Appl Physiol 83:810–816

Vernikos J (2020) Space and aging. In: Young LR, Sutton JP (eds) Handbook of bioastronautics. Springer, Cham. https://doi.org/10.1007/978-3-319-10152-1_96-3

Vernikos J, Schneider VS (2010) Space, gravity and the physiology of aging: parallel or convergent disciplines? A mini-review. Gerontology 56:157–166

Videbaek R, Norsk P (1997) Atrial distension in humans during microgravity induced by parabolic flight. J Appl Physiol 83:1862–1866

Voorhies AA, Ott CM, Mehta S, Pierson DL, Crucian BE, Feiveson A, Oubre CM, Torralba M, Moncera K, Zhang Y, Zurek E, Lorenzi HA (2019) Study of the impact of long-duration space missions at the International Space Station on the astronaut microbiome. Sci Rep 9:9911. https://doi.org/10.1038/s41598-019-46303-8

Wagner I, Braun M, Slenzka K, Posten C (2016) Photobioreactors in life support systems. Adv Biochem Eng Biotechnol 153:143–184. https://doi.org/10.1007/10_2015_327

Walsh L, Schneider U, Fogtman A, Kausch C, McKenna-Lawlor S, Narici L, Ngo-Anh J, Reitz G, Sabatier L, Santin G, Sihver L, Straube U, Weber U, Durante M (2019) Research plans in Europe for radiation health hazard assessment in exploratory space missions. Life Sci Space Res 21:73–82

Weber J, Javelle F, Klein T, Foitschik T, Crucian B, Schneider S, Abeln A (2019) Neurophysiological, neuropsychological, and cognitive effects of 30 days of isolation. Exp Brain Res 237 (6):1563–1573. https://doi.org/10.1007/s00221-019-05531-0

Wehland M, Grimm D (2017) Tissue engineering in microgravity. In: Ruyters G, Betzel C, Grimm D (eds) Biotechnology in space, SpringerBriefs in space life sciences. Springer, Cham. https://doi.org/10.1007/978-3-319-64054-9

Wheeler RM (2017) Agriculture for space: people and places paving the way. Open Agric 2:14–32

Wiederhold ML, Pedrozo HA, Harrison JL, Hejl R, Gao W (1997) Development of gravity-sensing organs in altered gravity conditions: opposite conclusions from an amphibian and a molluscan preparation. J Gravit Physiol 4:51–54

Yamaguchi NB, Roberts M, Castro S, Ozbre C, Makimura K, Leys N, Grohmann E, Sugita T, Ichijo T, Nasu M (2014) Microbial monitoring of crewed habitats in space – current status and future perspectives. Microbes Environ. https://doi.org/10.1264/jsme2.ME14031

Zeidel ML (2017) Salt and water: not so simple. J Clin Invest 127:1625–1626

Chapter 4
Success Stories: Innovative Developments for Biomedical Diagnostics and Preventative Health Care

Abstract Research and development programs such as space life sciences often require technical solutions for extreme environments and, in many cases, lead to improvements of existing or to the creation of new production methods or technologies, products, and instruments. Small, compact, light, easy-to-handle, and—if possible—non-invasive application—these are the characteristics of devices and methods, which astronauts prefer for physiological routine measurements and experiments. The same features however are also of great advantage on Earth, especially for newborns and elderly people as well as for routine application in hospitals and in extreme and isolated environments such as Antarctica or in submarines. Therefore, it is not too surprising that technologies and devices originally developed for space are routinely analyzed regarding the potential for terrestrial applications. Also, existing technologies and devices already available on Earth have been improved, tested, and applied in the harsh and challenging conditions of spaceflight thus promoting their commercial success. Overall, these achievements do not only stimulate new markets, industries, and opportunities but also improve the general quality of life. These benefits and the impact on science and economy however are often not immediate and obvious; and not seldom, it takes many years for a new idea to be transformed into a marketable product or service as will be shown in this chapter.

Keywords Biomedical diagnostics · Commercial devices · Countermeasures · Human physiology · Life cell imaging · Microbiome · Space medicine · Terrestrial application

4.1 Introduction

In Europe, especially through the activities led by ESA and DLR, the development of innovative experiment facilities and devices for biomedical space research has been and still is one of the strengths of the respective space life science programs. The leading role of Germany in human spaceflight over decades has ensured that DLR, German industry, and scientists are judged as reliable and interesting partners worldwide. By this, the program succeeded in providing major contributions to meet the global challenge of health and nutrition—as requested by the German government in its high-tech strategy document from 2010 (www.bmbf.de)—especially in the area of prevention of illness, its diagnosis, and therapy. In the following, examples for these developments including their successful commercial application will be presented with a focus on German achievements; some successes of other European countries, especially of France, have been summarized by Seibert (2001).

4.2   Cardiovascular System Diagnostics

As described in Chap. 3, microgravity leads to a rapid and substantial fluid shift from the lower to the upper part of the astronaut's body with consequences for the various components of the cardiovascular system including lung ventilation and intraocular pressure. Innovative, non-invasive devices have been developed to monitor these microgravity-induced changes, and several of them have found their way into the commercial market.

4.2.1   Self-Tonometer for Measuring Intraocular Pressure

For measuring intraocular pressure in astronauts non-invasively and as quick as possible after entering into microgravity, a small handheld self-tonometer was developed by scientists from the University of Hamburg in the late 1980s for the upcoming Spacelab and MIR missions (Groenhoff et al. 1992; Draeger et al. 1993). The tonometer is positioned by the subject after local anesthesia in front to the eye, a prism is then automatically advanced to contact and applanates the apex of the cornea. With the measured applanating force, the tonometer calculates the intraocular pressure.

After the successful application of the self-tonometer during the space missions MIR'92 and D-2 in 1992 and in 1993, respectively, a commercial version of the device was developed by the company EPSa (Elektronik und Präzisionsbau Saalfeld) in Jena, Germany, and introduced into the commercial market (Fig. 4.1). On Earth, self-tonometry is important because the intraocular pressure in glaucoma patients tends to fluctuate during the day so that repetitive measurements are necessary. The commercial device, called Ocuton S, allowed glaucoma patients to monitor their intraocular pressure themselves at home enabling the ophthalmologist to provide better care for the patient. In fact, experts stated that the use of Ocuton S appeared to be suitable to record diurnal profiles of intraocular pressure at home (Vogt and Duncker 2005). In 1999, Ocuton S won the Thuringia Future Prize.

A few years ago, Ocuton S got new attention in the context of a clinical research project called TT-MV (TeleTonometry Mecklenburg-Vorpommern). Several modifications were integrated, especially with regard to controlling the proper handling

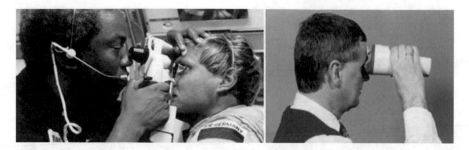

Fig. 4.1 Left: During the D1 Space Shuttle mission, NASA astronaut Guion S. Bluford checked the intraocular pressure of German ESA astronaut Ernst Messerschmid; for MIR'92 and D-2 a self-tonometer was developed, further improved for commercial use (right). (© NASA)

and positioning of the device now called Ocuton S*TT-MV (Lanfermann et al. 2009).

In a critical review of telemedicine technologies for diagnosis and monitoring of glaucoma, Strouthidis et al. (2014) attribute significant potential for telemetry-friendly devices to remotely monitor the progress of glaucoma. After further improvement of the efficacy and safety, these technologies for use by patients at home should lead to a significant reduction of the workload on the overstretched health service in the aging society.

4.2.2   Electrical Impedance Tomography (EIT) to Assess Lung Function: "Making Ventilation Visible"

The established clinical techniques used to investigate lung function all demonstrate certain disadvantages: either it is not possible to identify regional differences in lung function (multiple inert gas technique, for example) or the procedure involves exposing the patient to radiation (computer tomography, magnetic resonance tomography, scintigraphy) and requires extensive equipment. Scientists from Göttingen University succeeded in overcoming these disadvantages by applying electrical impedance tomography (EIT) thus making use of the electrical properties of cells. The signals are acquired using electrodes on a belt around the rib cage thereby measuring changes of electrical impedance in parallel to changes in aeration within the lungs. Following a complicated, yet quick, analysis based on image data processing, the regional lung aeration as well as the volume of blood and other liquids in the lungs and blood flow can be quickly established. The technique itself was already known for quite some time—the "new mathematics" was the key to success here. On parabolic airplane flights in 1999 and 2000, the new EIT system was tested (Fig. 4.2). For the first time, researchers were able to identify gravity-related regional changes in the function of the lungs as well as shifts in the volume of blood and liquid in the chest (Frerichs et al. 2001; see also Sect. 3.5.1.7).

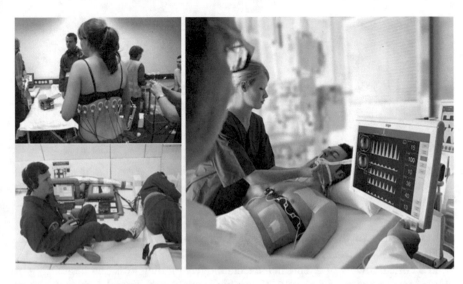

Fig. 4.2 Left: PulmoVista uses electrical impedance tomography to visualize and monitor breathing and lung fill. The mathematical methods to convert raw data to evaluable images were developed and successfully tested at changing gravity conditions during parabolic flights (© DLR); right: the commercial device (© Drägerwerk AG & Co. KGaA, Lübeck. All rights reserved)

Despite some experience with EIT utilization in research, the method had not been used in the routine clinical setting yet. The successful development of EIT at Göttingen University however led the company Drägerwerk AG & Co in Lübeck, Germany, to embark on this technique and to develop a commercial EIT device for clinical applications, especially for monitoring the lungs during artificial respiration. As a result of an intensive exchange of knowledge and experience between experts from Draeger company and the scientists from Göttingen university, the design, technique, and software were further improved. Since 2011, a commercial version— the so-called PulmoVista 500—is on the market with wide applications for medical diagnosis and monitoring of therapeutic maneuvers (Fig. 4.2). The advertising slogan "Making ventilation visible" nicely depicts the advantages of this new technology born from space research (www.draeger.com).

4.2.3   Measuring Body Core Temperature Noninvasively with the Double Sensor

As described in more detail in Sect. 3.5.1.6, a possibly dangerous increase in body core temperature of astronauts especially during exercise or extravehicular activities (EVAs) cannot be ruled out thus making the monitoring of the body core temperature a must. However, the measuring methods available such as rectal probes were

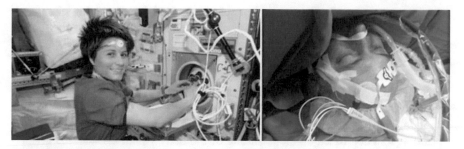

Fig. 4.3 The body core temperature of Italian ESA astronaut Samantha Christoforetti (left image) was recorded for the Circadian Rhythm experiment using the yellow thermosensor fixed to her forehead (© NASA/ESA). Another version of the sensor is used to monitor an intensive care patient. (Courtesy Prof. Gunga)

not suitable for astronauts and their daily routine. This is why a scientific team at the Charité Berlin developed a non-invasive measuring method in close cooperation with the Drägerwerk AG & Co. Sensors attached to the head and the breastbone measure heat flows which are then converted into body core temperatures with the aid of mathematical algorithms. The so-called double-twin sensor was patented in 2003 and 2006 (Gunga et al. 2008). After tests on parabolic flights, the thermosensor is being successfully used on the ISS since 2009 (Fig. 4.3).

The new thermosensor has great application potential especially for people in dangerous jobs, such as firefighters, police officers, and members of special duty forces; and, in addition, the entire discipline of occupational medicine benefits from non-invasive measurements. In clinical day-to-day work, the device is used in surgery as well as in neonate incubators. The new double-twin sensor has already been employed successfully in heart transplants carried out at the Berlin heart center. In 2015, the Drägerwerk company performed the market launch of the noninvasive thermosensor under the name Tcore advertising it with the slogan "Invasive methods are a thing of the past—new sensor helps reduce risk of hypothermia after surgery" (www.draeger.com). Recently, the thermosensor has successfully been combined with anesthesia devices. A further space application is again in preparation for ISS and future exploration missions beyond low-Earth orbit.

4.2.4 Ballistocardiography: An Old Method Revisited

Since many years, the cardiovascular system of cosmonauts on ISS is monitored in the projects "Pulse" and "Pneumocard," a successful Russian/German cooperation. Based on the space H/W commercial devices such as ECOSAN-7000 have been developed by the Russian partners (Ushakov et al. 2015; see also Sect. 3.5.1.3). Used in preventive medicine, for example, they help detect circulatory regulation disorders at an early stage. In sports medicine, they provide new ways to test an athlete's physical fitness, and in rehabilitation medicine to check the progress of patients with circulatory disorders. In the future, portable equipment for measuring circulatory functions, originally developed for use in space, will assist in monitoring the success of treatment in a patient's domestic environment (see also https://www.nasa.gov/mission_pages/station/research/news/b4h-3rd-ed-book).

For the follow-on space experiment called Cardiovektor, in which also scientists from Belgium are involved, an updated version of Pneumocard—capable of performing ballistocardiography (BCG)—was sent to the ISS in 2014. BCG is a non-invasive medical examination technique for assessing the mechanical efficiency of cardiac function based on the recording of the body vibrations with 3D accelerometers. Echocardiography, impedance cardiography, and respiration are providing physiological correlates of the 3D-BCG curve to interpret possible changes during the flight. Since 2015, Russian cosmonauts are participating in the Cardiovektor experiment. The scientists expect that the results from space will allow a better interpretation of the ballistocardiography measurements thus making a transfer of the technology towards medical or clinical applications on Earth possible (Luchitskaya et al. 2015; Migeotte et al. 2015).

While BCG had its glorious days in the 1960s and 1970s, it was nearly forgotten afterwards due to ECG, MRI, and other new methods, but also due to the technical difficulties and the size of the H/W required for the measurements. Currently, developments in the biomedical and engineering fields such as smaller and better accelerometers and electromechanical sensors as well as improved signal processing contribute to the idea of using ballistocardiography again in terrestrial settings. In fact, there is a regain of interest currently for BCG due to its potential for the non-invasive monitoring of elderly patients at home or for the diagnostic of pathologies affecting the cardiovascular system. Finally, BCG has the potential to provide crucial information on the mechanic cardiac function which no other technique can provide in an ambulatory, operator-independent way, possibly even embedded in office chairs and wheelchairs. In fact, under the leadership of the Belgian cooperation partners a respective spin-off device is under consideration.

4.2.5   Electronic Nose for Crew Health Monitoring via Breath Gas Analysis

Monitoring of crew health and performance including the detection of stress and of specific metabolic changes is of utmost importance in long-term exploratory missions, especially during heavy workload. In the past, blood samples often were taken, brought back to the ground, and analyzed. This is certainly not an option in long-term exploratory missions. To get quick and reliable information on the health status of astronauts during their spaceflight, breath gas analysis is a new approach.

The main contents of exhaled human breath are nitrogen, oxygen, water vapor, and carbon dioxide (in total 99.9%); but there are more than 200 other molecules, which are found as trace compounds. Since the release of certain bioproducts from

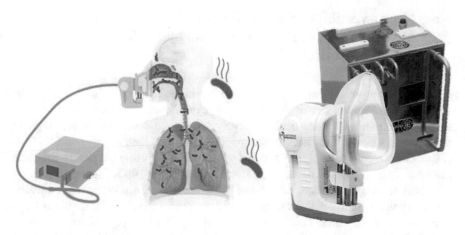

Fig. 4.4 The E-Nose Breath Gas Analyzer detects microbial volatile organic compounds as well as specific markers of pathological alterations, oxidative stress, and infections in the breath gas of astronauts. (© DLR)

tissue may reflect physiological processes or pathologic alterations such as oxidative stress, breath gas analysis may offer a significant diagnostic potential. Usually, mass spectrometry is applied for measuring trace gas compounds (Dolch et al. 2008, 2015), but for space, a new approach has been taken by DLR and scientists from LMU Munich. Based on the Electronic Nose concept originally developed by Airbus DS on contract by DLR for detection of microbial volatile organic compounds (MVOCs) that are emitted from microorganisms (see Sect. 4.6.1 for details), this technology is now adapted for breath gas analysis (Fig. 4.4). Here, specific markers in the expired breath gas, typical, e.g., for oxidative stress, are measured by the E-Nose upgraded with a newly developed breath gas collecting device.

The equipment was tested during a DLR parabolic flight campaign in October 2014. The E-Nose as well as the breath gas collecting device showed appropriate functionality and practicability (Dolch et al. 2017). On this basis, the equipment was prepared for its utilization on the ISS and launched in 2019 with the first breath gas analysis planned to be performed on Russian cosmonauts in the next few years.

Since biomarkers are known not only for oxidative stress but also for a variety of other diseases, such as lung cancer, inflammatory lung diseases, hepatic or renal dysfunction and diabetes, breath gas analysis by E-Nose may become an easy and quick non-invasive technology to get information on the health status of astronauts and of patients on Earth. Recently, the method has successfully passed a clinical test involving patients with pneumonia or lung failure, and is, in the frame of the COVID-19 pandemic already in a test phase for early detection of Coronavirus SARS-CoV-2 in breath gas. Results of ongoing experiments on parabolic flights and on the ISS will further support its development and application for astronauts as well as for critically ill people in intensive care. It is thus an interesting example for the successful adaptation and improvement of existing commercial technology leading to a benefit back for a variety of applications on Earth (Grosser et al. 2018).

4.3   Neuro-Vestibular System Diagnostics: Rapid Measurements of Eye Movements by the 3D-Eye Tracking Device

For investigating microgravity-induced changes in the functioning of the vestibular system via measurements of compensatory eye movements (see Sect. 3.5.2.1), several generations of video-oculography systems were developed in the frame of the German space life sciences program. VOG (Video-Oculography) and BIVOG (Binocular Video-Oculography) systems were successfully applied during Spacelab missions such as D-2 and in various MIR missions in the 1990s (Seibert 2001).

Based on this experience, scientists from Charité Berlin together with Kayser-Threde company in Munich developed a new system for the ISS in order to track eye movements in all three dimensions—the 3D-ETD (3D-Eye Tracking Device, Fig. 4.5). Several prizes have been awarded to the development, such as the "Innovation Prize on Clinical Information Technology 2000," the "Prize of the German Academy for Air and Travel Medicine 2000," and the "Hennig-Vertigo-Prize 2002" (for overview see Clarke 2017).

Based on the development of the space H/W, two start-up companies had already been founded in the 1990s out of Charité Berlin, namely ChronosVision (Berlin) and SMI (Sensomotoric Instruments; Teltow). SMI is mainly involved in providing the latest eye tracking solutions for applications ranging from medical research, diagnostics, and surgery, via psychology, performance sports, and man-machine interaction, up to market research (www.smivision.com).

ChronosVision is focussing especially on the scientific and clinical application including refractive laser surgery of the eyes and diagnostics of diseases such as virtual strabism. This is achieved by the commercial version, the Chronos Eye Tracker (C-ETD), as has been proven in numerous studies. Especially the application in refractive surgery (e.g., Lasik operation) for the correction of vision demands precise real-time measurement of eye position. Accordingly, throughout each laser

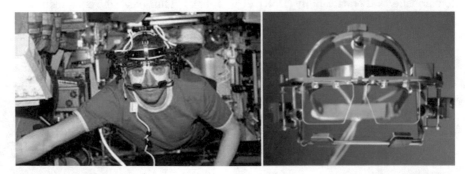

Fig. 4.5 Left: Russian cosmonaut Sergei K. Krikalev uses the 3D-Eye Tracking Device (ETD) in the Zvezda Service Module of the International Space Station to measure eye and head movements in space with great accuracy and precision (© ESA/NASA); right: example of commercial devices (ChronosVision)

operation, the position of the operated eye is measured by the video-oculographic eye tracker and relayed to the laser guidance control. As a consequence, the laser firing pattern is compensated for any changes in eye position. In April 2015, ChronosVision was inducted into the "Hall of Fame of Space Technology" for their sophisticated eye movement measurement equipment initially developed for experiments on the ISS (www.chronos-vision.de).

4.4 Bone and Muscle System Diagnostics

4.4.1 Backpain and Mobile Motion Analysis with Ultrasound

More than 50% of astronauts complain about back pain during their space missions. In trying to elucidate potential causes, a team from the German Sports University in Cologne started a project already in the 1990s to measure not only changes in spine length but also the movement behavior of astronauts in space. A respective space device for ultrasonic long-term monitoring of spine movements was developed in cooperation between scientists from the German Sports University and an engineering team from the University of Jena. The approach is based on the measurement of skin distances between four pairs of miniaturized ultrasound transmitters and receivers fixed on the back in parallel to the thoracic and lumbar spine (Fig. 4.6). For each channel, the skin distance between transmitter and receiver is determined at a sampling rate of 10 Hz. The transmitters and receivers are cable connected to a small data logger. Variations in skin distance represent a change in spine posture within the segment of the spine. The data points of the 12-channel system thus

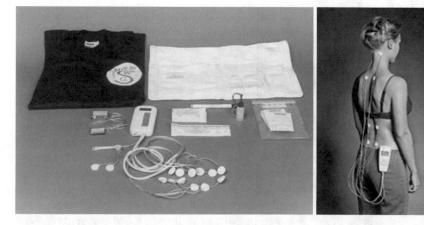

Fig. 4.6 The sonoSens hardware was flown on the MIR station in the late 1990s and used in bed rest studies as well as for occupational health analysis to collect spine movement data. (© DLR, Sensomotion.com)

represent a 3D model of the external spine curvature (Baum et al. 1997; see also Sect. 3.5.3.1).

After the successful utilization of the device in bed rest studies and on the MIR space station in the late 1990s, a commercial version was developed in cooperation with the company "friendly sensors" in Jena. Under the name sonoSens the commercial device was widely used for mobile motion analysis in orthopedics, ergonomics, automotive industry, occupational health, physiotherapy, and sports. Since 2004 the patented system is distributed in the USA, Germany, Austria, Japan, and Korea, today by the company Agentur-Graupner.

More recently, scientists from the German Sports University and the University of Cologne as well as from the Central Institute of the Federal Armed Forces Medical Services in Koblenz, Germany, have reported on a new approach to assess movements and isometric postures based on sonoSens. With a novel data analyzing software—the so-called JSpinal, especially with the implementation of algorithms to derive and calculate magnitude, duration, direction, and velocity of every single spine and trunk motion they were able not only to measure the motion continuously but also come to recommendations of individual preventive strategies and objective evaluation of ergonomics (Wunderlich et al. 2011).

In the context of occupational health studies, the so-called 3D-SpineMoveGuard (3D-SMG) was successfully tested and applied in different working environments such as with medical doctors during surgery or in dentistry, workers harvesting vegetables, in the construction sector, and in helicopter pilots (see, e.g., Wunderlich et al. 2010a, b). Considerations on commercialization of this updated 3D-SMG version of sonoSens are being taken. Since NASA quite recently identified space adaptation back pain as a major risk for exploration missions, interesting perspectives for this device for space application may open up again. Also, for future explorative missions, an international group of scientists is working on solutions to implement an ultrasound device onboard of the Lunar Gateway, which is part of NASA's Artemis Program.

4.4.2 Noninvasive Measurement of Muscle Properties by Digital Palpation

To assess changes in muscle properties of astronauts, a new noninvasive device, MyotonPro, has been developed by the company Myoton in Estonia and space qualified by OHB (Bremen). In the frame of an ESA/DLR project, it currently supports—together with ultrasound imaging—the ISS experiment Myotones, proposed by an international team and led by scientists from the Charité Berlin. Previously, the device had been tested successfully in various clinical settings and in an ESA parabolic flight campaign in 2015 as well as in a bed rest study at DLR Cologne with reactive jumps as a countermeasure (Schneider et al. 2015; Schoenrock et al. 2018).

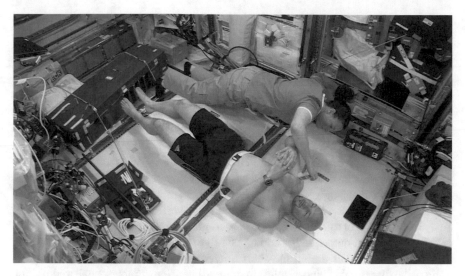

Fig. 4.7 US astronaut Serena Auñón-Chancellor uses MyotonPRO to measure tone, tension, and stiffness of the muscles of German ESA astronaut Alexander Gerst on ISS. (© ESA/NASA)

MyotonPro is a small device of the size of a mobile phone to measure non-invasively key biomechanical parameters of the superficial skeletal muscle (Fig. 4.7). Parameters include tone, tension, stiffness, elasticity, relaxation, and relate to the relaxed skeletal muscle. The method of measurement—the digital palpation method—consists of recording damped natural oscillation of soft biological tissue in the form of an acceleration signal and the subsequent simultaneous computation of the parameters of state of tension, biomechanical, and viscoelastic properties. Damped natural oscillation is induced by an exterior, low force quick-release mechanical impulse under constant preload (www.myoton.com).

In future, MyotonPro will yield a more comprehensive view of the fitness of astronauts and of the efficacy of countermeasure programs. Thus, it will be very useful for long-term health monitoring of astronauts during exploration missions beyond low-Earth orbit. On Earth, it is used for new approaches in neurorehabilitation, orthopedics, muscle dystrophy, and aging research in general. It also has great potential for monitoring and assessing fitness and training programs of athletes.

4.5   Live Cell Imaging with FLUMIAS

FLUMIAS (Fluorescence Microscopy Analyses in Space) is a high-resolution fluorescence microscope for live cell imaging on microgravity platforms. It provides the capabilities to take a close look inside cells of the human body and other organisms from single cells and microorganisms to plants and animals with high temporal and spatial resolution. Changes in metabolic processes, membrane dynamics, and ion flows caused—for instance—by varying gravity conditions can be viewed in real-time. Structural changes in internal supporting and organizational structures of the cells, such as the cytoskeleton, are made visible. Protein structures in human immune cells are labeled with fluorescing markers and made visible using the FLUMIAS microscope (see also Sect. 3.5.4.5). However, the FLUMIAS application is certainly not limited to biology. The amazing capabilities have also attracted the attention of international physical and material science teams interested in studying gravity-dependent processes on the microscopical level on ISS.

The development of FLUMIAS is being undertaken by DLR in several steps. At first, spinning-disc microscope models were developed for experiments on DLR parabolic flight campaigns as well as for TEXUS sounding rocket flights on contracts to Airbus DS. It was the first time, scientists were able to directly observe and analyze the effects of microgravity and altered gravity on the structure and dynamics of the cytoskeleton mitochondria and other cell organelles in human immune, nerve, and breast cancer cells during the short microgravity phases of TEXUS and parabolic plane flights using high-resolution, three-dimensional fluorescence microscope images. In fact, the understanding of the fundamental role of gravity in cancer cell growth and function is a new paradigm in cell biology obtained from experiments performed in microgravity (Corydon et al. 2015; Krüger et al. 2017).

The successful utilization of these flight opportunities led to the decision to develop the so-called FLUMIAS ISS technology demonstrator, a simplified version of the planned space station device using "structured illumination technology" for the generation of high-resolution 3D fluorescence images. The demonstrator was launched to ISS with Space-X CRS 15 end of June 2018 and was tested during the Horizons mission of Alexander Gerst (Fig. 4.8). The excellent-quality live cell images that were sent to Earth along with perfect technical data confirmed the space suitability of the new high-resolution live cell imaging fluorescence microscope technology and identified some critical components (Carstens et al. 2018).

The development of the ISS flight model of FLUMIAS is currently underway. Mounted on a centrifuge rotor, this compact version will allow the examination of scientific samples by live cell imaging under acceleration conditions between microgravity and 1 g. Above all, directly visualizing the effects of microgravity and different gravitational forces on biological, biochemical, and physical processes is expected to lay the foundation for numerous scientific breakthroughs in a wide range of research areas in the near future.

4.6 Technologies for Fighting Microbial Challenges

4.6.1 Electronic Nose for Microbial Monitoring

As pointed out in Sect. 3.3.3, the growth of microorganisms and fungi can pose a major threat not only for the health of astronauts but can also cause corrosion of

Fig. 4.8 US astronaut Drew Feustel preparing the FLUMIAS technology demonstrator for installation in Space Tango's TangoLab on ISS with ESA astronaut Alexander Gerst in the background. (© ESA/NASA)

material in long-term spaceflight. Especially behind facility panels and racks, microorganisms were found in large amounts in the Russian MIR station, but also on the ISS, deteriorating materials and potentially posing a health risk to humans onboard. In the past, samples—for instance—from contaminated surfaces or filters were collected and sent back to Earth for subsequent analysis. For long-term exploratory missions, it is obvious however that an onboard detection and analysis is needed to enable countermeasures in time.

Several so-called electronic noses have already been developed for different purposes (for overview see Reidt et al. 2017). On the ISS, an Electronic Nose system is available developed by the Jet Propulsion Laboratory that is utilized for the detection of certain gaseous components in air, but not those emitted from microorganisms. The so-called MonoNose was used for the identification of microorganisms in clinical samples. Recently, a so-called Electronic Nose (E-Nose) has been developed by Airbus on contract by the German Space Agency DLR. Different sensors measure a characteristic odor signature of the probed sample, so-called microbial volatile organic compounds (MVOCs), which are emitted by microbes and allow for microbe-specific identification.

In the frame of a technical demonstration, first experiments have been performed in cooperation with Russian scientists in the Zvezda module in 2012/2013 showing the successful application of the E-Nose device for detection of biocontamination. Further development of the E-Nose system will involve a massive expansion of microbial training (Reidt et al. 2017).

4.6.2 AgXX®: An Innovative Antimicrobial Surface Technology

Microbial contamination, especially the formation of biofilms consisting of bacteria, yeast, or fungi pose a serious threat to space crews and the entire space infrastructure. On the one hand, they may impair the astronauts' health, while on the other hand, hardware may be damaged because of their corrosive impact on a wide range of materials. On Earth, the increase in the number of multiresistant pathogens requires alternative agents to antibiotics to avoid microbial infections. Also, here, biofilms cause problems in all fields associated with water quality, aqueous solutions, and hygiene, i.e., in hospitals, old people's homes, childcare facilities, in food technology, production, air conditioning, and also in households.

The use of silver-coated surfaces or silver nanoparticles is an option. However, many bacteria have developed silver resistance, and the cytotoxicity of nanoparticles causes problems in medical therapy. A few years ago, the interdisciplinary research of surface technologists, chemists, process engineers, and microbiologists involved also in space projects has led to the development of an entirely new antimicrobial contact catalyst, AgXX®, by Largentec GmbH (today AgXX® Laboratories Largentec GmbH). After initial commercial applications in the field of microbial contamination of aqueous solutions and processing waters, tests on the ISS were considered and implemented by an international team consisting of scientists from Freiburg University, Beuth University of Applied Sciences, Berlin, and AgXX® company as well as from the IBMP Moscow.

For the ISS experiment conducted between 2012 and 2015, specimen plates were prepared consisting of different carriers (stainless steel, polydimethylsiloxane) and coated with silver or AgXX® and analyzed for bacterial growth after return to Earth. Results were promising: the AgXX®-coated stainless steel specimen carriers were found free from microbial contamination after 6 months, only slight microbial growth could be detected after 12 or 19 months in space, while the other carriers were extensively infested by microbes. Moreover, it was found that the few microorganisms on the AgXX®-coated stainless steel specimen carriers were incapable of propagating thus presenting no threat to astronauts or material. Very recently, the scientists have postulated an action mechanism: obviously, the combination of silver ions and reactive oxygen species generated by AgXX® results in a synergistic antimicrobial effect, superior to that of conventional silver coatings (Clauss-Lendzian et al. 2018; Sobisch et al. 2019).

In total, AgXX® was successfully tested against more than 130 microorganisms, including superbugs on ISS. Since the AgXX® technology works independently and virtually without the release of toxic materials such as silver ions and a variety of different materials can be coated, such as metal, glass, or various polymer plastics, the application of AgXX® in space as well as on Earth is widespread. Food industry, healthcare, biomedicine, and drinking water conservation are just a few examples.

4.7 Countermeasure Devices: Maintaining Health and Performance in Astronauts and Humans on Earth in the Aging Society

It is well-known that the microgravity conditions of space have a negative impact on various systems of the human body. Cardiovascular deconditioning and neuro-vestibular problems, deterioration of the immune system as well as muscle and bone loss are frequently experienced by astronauts. To maintain crew health during long-term space missions, adequate countermeasures are necessary (for an overview see Hargens et al. 2013). Ideally, an exercise-based countermeasure should preserve the integrity and function of bones, neuromuscular, and cardiovascular system. In addition, the countermeasure should be efficient, requiring a minimum of crew time. Currently, the countermeasures applied on the ISS are far from being optimal: astronauts onboard the ISS are required to exercise about 2 h per day using at least three different countermeasure systems—a treadmill, a cycle ergometer, and the advanced resistive exercise device ARED. Nevertheless—while the cardiovascular system seems to remain in good shape—this time-consuming exercise is not successful in fully maintaining muscle and bone mass and function. Space agencies and scientists around the world are still looking for more efficient and integrated countermeasures. Also, ESA and DLR have gathered some expertise in the development and utilization of different countermeasure systems. This is especially due to their application in several bed rest studies, often jointly performed by ESA, CNES, and DLR in Toulouse, Cologne, and Berlin.

4.7.1 Lower Body Negative Pressure Device for Stimulating Blood Flow

A microgravity-induced fluid shift from the lower to the upper parts of the body is well-known to be one of the earliest physiological changes that astronauts experience in spaceflights. In the past, various devices have been developed and applied that provide low pressure to the lower body (Lower Body Negative Pressure Device, LBNP) to counteract this fluid shift, especially as preparation for the landing of astronauts (for review see Campbell and Charles 2015). A foldable device providing the possibility to perform microneurography in space via zippers in the LBNP was developed by the German space company ASTRIUM (Friedrichshafen; today Airbus Defence and Space) for the space shuttle mission NEUROLAB in 1998 (Fig. 4.9). The development was based on an idea of scientists from the DLR Institute of Aerospace Medicine in Cologne. The "NEUROLAB LBNP" was successfully used for experiments by the so-called Autonomous Nervous Function Team, scientists responsible for a suite of four experiments in the area of blood pressure control (Baisch et al. 2003; Buckey and Homick 2003).

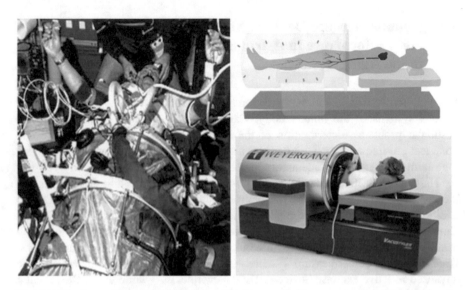

Fig. 4.9 The Lower Body Negative Pressure Device was used in space to reduce blood shift to the head of astronauts in microgravity. The device has evolved into a commercial product that, for example, helps patients with cardiovascular disorders on Earth to increase blood/oxygen supply to the leg vessels. (© NASA, DLR, Weyergans High Care AG)

Even though an already planned follow-on project for ISS was never realized, the ideas of the DLR Institute of Aerospace Medicine for the Neurolab and ISS LBNP led the company Weyergans High Care (Düren, Germany) to develop several versions of a commercial device for application in health care. As Vacumed, Vacustyler, or Vacufit the devices are widely used for the treatment of circulatory problems, cellulitis, lymph drainage, and for improvement of the physiological muscle pump. The patented application method—today known as IVT (Intermittent Vacuum Therapy)—is carried out worldwide in leading institutes for curative cosmetics, medical wellness, and health and prevention, but also in clinics and rehabilitation centers or sports clubs according to the slogan of Weyergans "innovative cosmetics, medicine and fitness concepts based on space medicine proven on Earth" (for further details see www.weyergans.de).

In preparation for exploratory missions, the LBNP concept is currently experiencing a revival. Due to the confined space in the Gateway and other space vehicles for long-term travel that do not allow bulky training equipment, LBNP is being reconsidered as an important and effective countermeasure device preventing numerous adverse effects of microgravity.

4.7.2 Vibration Training with Galileo

4.7.2.1 Improvement of Muscle and Bone Health in Astronauts and in Elderly People and Patients on Earth

Confinement to bed, lack of exercise, and paralysis as well as wearing a plaster cast, and, not least, living in microgravity inevitably leads to muscular atrophy and bone loss after a few weeks. To keep the muscles and bones of astronauts on long-term missions as "normal" as possible, a training device named "Galileo Space" was developed by Novotec which basically consists of a side-alternating vibrating plate (www.galileo-training.com). The device provides sinusoidal, repetitive motion, resulting in motor and neural learning effects. The side-alternating vibration simulates a physiologically meaningful movement pattern similar to human gait, which makes it possible to effectively train the back muscles. In 2003/2004 a bed rest study in which subjects were confined to bed for 8 weeks showed that those who had to press against this vibrating plate for a few minutes each day while lying in bed lost only about 10% of their muscular strength and 0.5% of their bone density. Subjects without such training, on the other hand, lost 40% of their muscular strength and 4.5% of their bone density (Blottner et al. 2006; Belavý et al. 2009, 2011).

Tests of using the vibration device under varying gravity conditions have been successfully performed in DLR parabolic flight experiments a few years ago (Fig. 4.10). Since the duration of exercise can be drastically reduced to a couple of minutes per day when applying vibration training, the method is still being discussed among ISS partners as a new countermeasure for astronauts, maybe in combination with other training methods such as reactive jumps (Ritzmann et al. 2014 and see below).

On Earth, Galileo vibration training is meanwhile used for several years in a variety of different applications, from fitness clubs to rehabilitation and in the clinic. Under the headline "Auf die Beine" ("On your legs"), a new therapy concept for

Fig. 4.10 The Galileo vibration platform, developed to mitigate muscle and bone loss in space, was tested during parabolic flights. Commercial versions of Galileo help elderly people and patients to improve muscle and bone health and are increasingly used for fitness training. (DLR, Novotec Medical GmbH, Pforzheim)

children with various diseases, e.g., osteogenesis imperfecta, was launched at the university hospital Cologne as a joint effort of Prof. Schönau, BEK health insurance company and medifitreha (for further publications see unireha.uk-koeln.de). At the children's hospital of Cologne, for instance, the device is successfully being used to treat children suffering from brittle-bone disease. Also, elderly people, patients after surgery, or with thrombosis benefit from the vibration training (for review see Runge 2006; Fig. 4.10). An impressive summary of publications describing the successful application of vibration training with Galileo is listed under https://www.galileo-training.com/de-english/literature/galileo-publikationen.html.

4.7.2.2 Vibration Training for Patients with Cerebral Palsy

Studies on patients confined to bed rest clearly document that vibration training effectively counteracts muscle and bone attenuation. But what are the neurophysiological mechanisms that are causing this effect? This was being investigated by scientists at Freiburg University in a project that aimed to uncover the interaction between nerve and muscle. They were able to show that vibration training greatly reduces spinal excitability which however returns to normal after about 10 min. Also, this result was not affected by the changing gravity conditions of a parabolic flight confirming that the method is suitable for use by astronauts in microgravity (Kramer et al. 2013; Ritzmann et al. 2013).

Based on these discoveries, vibration training found its way into a special clinical rehabilitation. In humans with cerebral palsy, everyday motor skills like walking or standing are greatly impaired because they suffer from increased reflex activity. Now that this spinal excitability can be reduced effectively by vibration training, cerebral palsy may be treated more successfully. In fact, patients improved their locomotion as well as their coordination and fitness abilities. Thus, as concluded in a recent systematic review on the effectiveness of vibration therapy, its acute and chronic application as a nonpharmacological approach has the potential to ameliorate symptoms of cerebral palsy, achieving significant improvements in daily living (Ritzmann et al. 2018).

4.7.3 The Sledge Jump System: An Efficient Tool to Preserve Fitness Within Minutes

Looking for more effective and integrated countermeasures, one type of exercise has emerged in the last few years that seems to combine several advantages: peak load jumping. Jumping is a high-intensity, low-volume type of training that does not require much time, yet reactive jumping induces high strain and strain rates, which

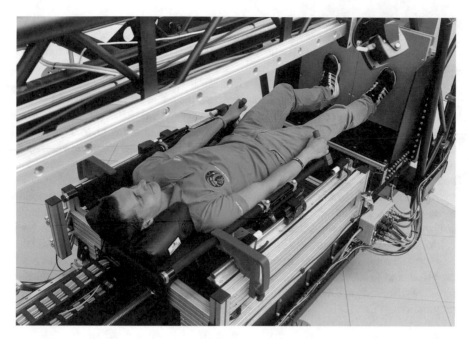

Fig. 4.11 The sledge jump system mounted on the :envihab short-arm human centrifuge was tested as countermeasure device during a bed rest study at DLR Cologne. (© DLR, all rights reserved)

have been found to be key determinants for bone strength. In addition, jump training increases leg muscle strength and even counteracts cardiovascular deconditioning.

For application in bed rest studies, a sledge jump system was developed by Novotec Medical GmbH in Pforzheim and scientists from Freiburg and Konstanz Universities under an ESA contract with Airbus. It consists of a frame on wheels and a lightweight sledge (5 kg) that is attached to a rail on both sides of the frame. The construction allows the sledge only to slide alongside the rails. The subject is attached to the sledge via two straps around the shoulders allowing movements in a natural manner. Four low-pressure cylinders generate the force that pulls the sledge towards the force plates. By altering the pressure of the cylinders, forces between zero and 1800 N can be set (Kramer et al. 2017a). The original design even allows to tilt the table continuously for more than 90° from the vertical, which enables training under reduced gravity such as on Moon or Mars. Also jumps under hypergravity are feasible by an increase of the pullback force (Fig. 4.11).

In a recent 60-day bed rest study in Cologne, the sledge jump system was evaluated as a countermeasure. These results were very convincing: approximately 3 min of jump training per session with 5–6 sessions per week turned out to be sufficient not only to maintain high peak force and even to increase jump height and power but also to preserve lean body mass and to prevent cardiovascular deconditioning. Jump training prevented bed rest-induced atrophy, as indicated by parameters like slow-to-fast myofiber shift and reduced myofiber cross-sectional area in healthy male subjects (Blottner et al. 2020). The training was well tolerated and accepted by the test subjects, also from the psychological point of view as measured by the so-called Physical Activity Enjoyment Scale. It thus seems that jumping training with the sledge jump system has the potential to fulfill all necessary requirements for a countermeasure that leads to a very time-efficient and effective type of exercise for astronauts on long-term missions (Kramer et al. 2017b). More than 2 h of training per day—with different devices, such as bicycle ergometer, treadmill, and ARED, as currently required from the ISS astronauts—would then belong to history.

Application on Earth is straightforward: The development of a training program to be performed within a few minutes per day seems like a dream for all people striving to maintain or even improve their performance and physical fitness. It also contributes to counteracting the consequences of inactivity in the aging and sedentary populations such as osteoporosis, sarcopenia, diabetes, and cardiovascular problems and to preventing the deterioration of neuromuscular performance during physical inactivity (Kramer et al. 2018). Respective clinical trials are under preparation.

4.8 Innovative Developments and Commercial Success

As demonstrated in the previous sections, quite a number of innovative developments for biomedical diagnostics and healthcare in space have found their ways into the commercial market on Earth. This is especially true for small devices monitoring noninvasively parameters of the cardiovascular, cardiopulmonary, the vestibular, and the musculoskeletal systems. Also, countermeasure devices developed for or validated in space have contributed here. Table 4.1 provides a summary of these success stories of the German life sciences program also indicating the potential of some newly developed equipment. Additional information has been compiled by Seibert (2001) for Europe and very recently by Sutton (2020) mainly for the US and can also be found on the respective websites of ESA and NASA. Especially small and medium enterprises (SMEs) stand to benefit here, some of them making a large share of their revenues from these space-driven developments. It can be expected that in the era of space exploration new success stories of the life sciences programs worldwide can be written in upcoming years.

Table 4.1 Developments for biomedical diagnosis, monitoring, and countermeasure

Biomedical device/ technology	Parameter measured/ purpose	Space application	Commercial device
Self-tonometer	Intraocular pressure	MIR'92, D-2	Ocuton S
Electrical impedance tomography	Lung function and ventilation	Parabolic Flights	Pulmovista 500
Pulse/Pneumocard	Various cardiovascular parameters	ISS Russian Module	Ecosan 7000
Cardiovektor/ Ballistocardiography	Cardiac function	ISS Russian Module	In development
Thermosensor	Body core temperature	ISS	Tcore™
Electronic nose for breath gas analysis	Health monitoring	ISS	
3D-eye tracking device	Eye movement	ISS	Various eye trackers
Motion measurement device	Spine geometry; mobile motion analysis	MIR	sonoSens
Digital palpation	Various muscle properties	ISS	MyotonPro
FLUMIAS	Live cell imaging with rotating disc	PF, TEXUS, ISS	
Electronic nose	Microbial monitoring	ISS Russian Module	
Antimicrobial surface technology	Antimicrobial surface coating	ISS Russian Module	AgXX®
Lower body negative pressure	Underpressure treatment	Spacelab	Vacumed, Vacustyler
Vibration device	Vibration training	Bed rest studies, Parabolic Flights	Galileo
Sledge jump system	Jumping training	Bed rest studies	

References

Baisch FJ, Beck EJ, Gerzer R, Moller K, Wesseling KH, Drummer C, Heer M, Karemaker JM (2003) Blood pooling and plasma filtration in the thigh in microgravity. In: Buckey JC, Homick JL (eds) The Neurolab Mission: neuroscience research in space. NASA SP-2003-535. Johnson Space Center, Houston, TX, pp 203–205

Baum K, Hoy S, Essfeld D (1997) Continuous monitoring of spine geometry: a new approach to study back pain in space. Int J Sports Med 18:331–333

Belavý D, Miokovic T, Armbrecht G, Rittweger J, Felsenberg D (2009) Resistive vibration exercise reduces lower limb muscle atrophy during 56-day bed rest. J Musculoskelet Neuronal Interact 9:225–235

Belavý D, Beller G, Armbrecht G, Perschel FH, Fitzner R, Bock O, Börst H, Degner C, Gast U, Felsenberg D (2011) Evidence for an additional effect of whole-body vibration above resistive exercise alone in preventing bone loss during prolonged bed rest. Osteoporosis Int 22:1581–1589

Blottner D, Salanova M, Püttmann B, Schiffl G, Felsenberg D, Buehring B (2006) Human skeletal muscle structure and function preserved by vibration muscle exercise following 55 days of bed rest. Eur J Appl Physiol 97:261–271

Blottner D, Hastermann M, Weber R, Lenz R, Gambara G, Limper U, Rittweger J, Bosutti A, Degens H, Salanova M (2020) Reactive jumps preserve skeletal muscle structure, phenotype, and myofiber oxidative capacity in bed rest. Front Physiol 10:1527. https://doi.org/10.3389/fphys.2019.01527

Buckey JC, Homick JL (eds) (2003) The Neurolab mission: Neuroscience research in space. NASA SP-2003-535. Johnson Space Center, Houston, TX, pp 171–205

Campbell MR, Charles JB (2015) Historical review of lower body negative pressure research in space medicine. Aerospace Med Hum Perform 86:633–640

Carstens AC, Braun M, Ullrich O, Thiel C, Treichel R (2018) FLUMIAS demonstrator: a miniature fast-track approach to live cell imaging microscopy on the ISS. 69th IAC Bremen. IAC-18-A.1.8.14

Clarke AH (2017) Vestibulo-oculomotor research in space, SpringerBriefs in space life sciences. Springer, Cham. https://doi.org/10.1007/978-3-319-59933-5

Clauss-Lendzian E, Vaishampayan A, de Jong A, Landau U, Meyer C, Kok J, Grohmann E (2018) Stress response of a clinical Enterococcus faecalis isolate subjected to a novel antimicrobial surface coating. Microbiol Res 207:53–64

Corydon TJ, Kopp S, Wehland M, Braun M, Schütte A, Mayer T, Hülsing T, Schmitz B, Hemmersbach R, Grimm D (2015) Alterations of the cytoskeleton in human cells in space proved by life-cell imaging. Nat Sci Rep 6:20043. https://doi.org/10.1038/srep20043

Dolch ME, Frey L, Hornuss C, Schmoelz M, Praun S, Villinger J, Schelling G (2008) Molecular breath gas analysis by online mass spectrometry in mechanically ventilated patients: a new software-based method of CO_2-controlled alveolar gas monitoring. J Breath Res 2:037010

Dolch ME, Choukèr A, Hornuss C, Frey L, Irlbeck M, Praun S, Leidlmair C, Villinger J, Schelling G (2015) Quantification of propionaldehyde in breath of patients after lung transplantation. Free Radic Biol Med 85:157–164

Dolch ME, Hummel T, Fetter V, Helwig A, Schelling G (2017) Electronic nose, functionality for breath gas analysis during parabolic flight. Microgr Sci Technol 29:201–207

Draeger J, Schwartz R, Groenhoff S, Stern C (1993) Self-tonometry under microgravity conditions. Clin Investig 71:700–703

Frerichs I, Dudykevych T, Hinz J, Bodenstein M, Hahn G, Hellige G (2001) Gravity effects on regional lung ventilation determined by functional EIT during parabolic flights. J Appl Physiol 91:39–50

Groenhoff S, Draeger J, Deutsch C (1992) Self-tonometry: technical aspects of calibration and clinical application. Int Ophthalmol 16:299–303

Grosser J, Lenic J, Kharin S, Tsarkov D, Smirnov Y, Novikova N, Moukhamedieva L, Dolch M, Fetter V, Hummel T, Reidt U, Kornienko A, Nitzer R, Helwig A, Roth P (2018) E-Nose: measuring surface microbial contamination and oxidative stress of cosmonauts – results and future application. 69th IAC Bremen. IAC-18, A1,7,8

Gunga HC, Sandsund M, Reinertsen RE, Sattler F, Koch J (2008) A non-invasive device to continuously determine heat strain in humans. J Therm Biol 33:297–307

Hargens AR, Bhattacharya R, Schneider SM (2013) Space physiology VI: exercise, artificial gravity, and countermeasure development for prolonged space flight. Eur J Appl Physiol 113:2183–2192

Kramer A, Gollhofer A, Ritzmann R (2013) Acute exposure to microgravity does not influence the H-reflex with or without whole body vibration and does not cause vibration-specific changes in muscular activity. J Electromyogr Kinesiol 23:872–878

Kramer A, Kümmel J, Mulder E, Gollhofer A, Frings-Meuthen P, Gruber M (2017a) High-intensity jump training is tolerated during 60 days of bed rest and is very effective in preserving leg power and lean body mass: an overview of the Cologne RSL study. PLoS One. https://doi.org/10.1371/journal.pone.0169793

Kramer A, Gollhofer A, Armbrecht G, Felsenberg D, Gruber M (2017b) How to prevent the detrimental effects of two months of bed-rest on muscle, bone and cardiovascular system: an RCT. Sci Rep. https://doi.org/10.1038/s41598-017-13659-8

Kramer A, Kümmel J, Gollhofer A, Armbrecht G, Ritzmann R, Belavy D, Felsenberg D, Gruber M (2018) Plyometrics can preserve peak power during 2 months of physical inactivity: an RCT including a one-year follow-up. Front Physiol. https://doi.org/10.3389/fphys.2018.00633

Krüger M, Wehland M, Kopp S, Corydon TJ, Infanger M, Grimm D (2017) Life-cell imaging of F-actin changes induced by 6 min of microgravity on a TEXUS sounding rocket flight. In: Proc. 23rd ESA symposium on European rocket and balloon programmes and related research

Lanfermann E, Jürgens C, Großjohann R, Antal S, Tost F (2009) Intraocular pressure measurements with the newly reconfigured Ocuton S*TT-MV self-tonometer in comparison to Goldmann applanation tonometry in glaucoma patients. Med Sci Monit 15:556–562

Luchitskaya E, Baevsky R, Funtova I, Tank J (2015) Further development of the inflight experiment Cardiovektor. IAC-15,A1,3,3,x29177.brief.pdf

Migeotte PF, Funtova I, Baevsky RM, Prisk GK, Tank J, Limper U, Möstl S, Gauger P, Lejeune L, Deliere Q, Van De Borne P, Schlegel H, Luchitskaya E (2015) Three-dimensional ballistocardiography in microgravity at 20 years interval: a longitudinal case report and validation of the esa-b3d project. In: Proceedings of the International Astronautical Congress. pp 85–90

Reidt U, Helwig A, Müller G, Plobner L, Lugmayr V, Kharin S, Smirnov Y, Novikova N, Lenic J, Fetter V, Hummel T (2017) Detection of microorganisms onboard the International Space Station using an electronic nose. Gravit Space Res 5:89–111

Ritzmann R, Gollhofer A, Kramer A (2013) The influence of vibration type, frequency, body position and additional load on the neuromuscular activity during whole body vibration. Eur J Appl Physiol 113:1–11

Ritzmann R, Freyler K, Krause A, Gollhofer A (2014) Auswirkung von Schwerelosigkeit auf den menschlichen Bewegungsapparat. Flug- und Reisemedizin 21:176–182

Ritzmann R, Stark C, Krause A (2018) Vibration therapy in patients with cerebral palsy: a systematic review. Neuropsychiatr Dis Treat 14:1607–1625

Runge M (2006) Die Vibrationsbehandlung – neue Wege in Therapie und Training. Bewegungstherapie und Gesundheitssport 22:1–5

Schneider S, Peipsi A, Stokes M, Knicker A, Abeln V (2015) Feasibility of monitoring muscle health in microgravity environments using Myoton technology. Med Biol Eng Comput 53:57–66. https://doi.org/10.1007/s11517-014-1211-5

Schoenrock B, Zander V, Dern S, Limper U, Mulder E, Veraksitš A, Viir R, Kramer A, Stokes MJ, Salanova M, Peipsi A, Blottner D (2018) Bed rest, exercise countermeasure and reconditioning effects on the human resting muscle tone system. Front Physiol 9:810. https://doi.org/10.3389/fphys.2018.00810

Seibert G (2001) A world without gravity. ESA SP-1251. pp 317–318

Sobisch LY, Rogowski KM, Fuchs J, Schmieder W, Vaishampayan A, Oles P, Novikova N, Grohmann E (2019) Biofilm forming antibiotic resistant gram-positive pathogens isolated from surfaces on the International Space Station. Front Microbiol. https://doi.org/10.3389/fmicb.2019.00543

Strouthidis NG, Chandrasekharan G, Diamond JP, Murdoch IE (2014) Teleglaucoma: ready to go? Br J Ophthalmol 98:1605–1611

Sutton JP (2020) Clinical benefits of bioastronautics. In: Young LR, Sutton JP (eds) Handbook of bioastronautics. Springer, Cham. https://doi.org/10.1007/978-3-319-10152-1_79-2

Ushakov IB, Orlov OI, Baevskii RM, Bersenev EY, Chernikova AG (2015) New technologies for assessing health in essentially healthy people. Neurosci Behav Physiol 45:116–120. https://doi.org/10.1007/s11055-014-0047-7

Vogt R, Duncker GIW (2005) Anwendbarkeit der Selbsttonometrie unter ambulanten Bedingungen zur Gewinnung von Tagesprofilen des intraokulären Druckes. Kln Monatsbl Augenheilk 222:814–821

Wunderlich M, Eger T, Rüther T, Meyer-Falcke A, Leyk D (2010a) Analysis of spine loads in dentistry – impact of altered sitting position of the dentist. J Biomed Sci Eng 3:664–671

Wunderlich M, Jacob R, Rüther T, Leyk D (2010b) Analysis of spinal stress during surgery in otolaryngology. HNO 58:791–798

Wunderlich M, Rüther T, Essfeld D, Erren TC, Piekarski C, Leyk D (2011) A new approach to assess movements and isometric postures of spine and trunk at the workplace. Eur Spine J 20:1393–1402

Chapter 5
Space Life Sciences in the Exploration Era: An Outlook on Future Challenges and Opportunities

The Earth is the cradle of humanity,
but mankind cannot stay in the cradle forever.
—Konstantin Tsiolkovsky

Abstract After many years of space life science research performed on occasional flight opportunities and more than 20 years of continuous and constantly increasing utilization of the International Space Station, we are now at the brink of a new era. Human space exploration will make extensive use of a multitude of scientific results, instruments, and experience, but will undoubtedly still require a lot more preparation activities to close the gaps in knowledge and technology before we can send humans to destinations beyond low-Earth orbit. This chapter discusses future priorities, new approaches, developments, and technologies for exploration-oriented life sciences.

Keywords Artificial intelligence · Exploration technologies · Human space exploration · Exploration destination · Space life science priorities · Telemedicine

In a certain way, the docking of the Apollo and Soyuz spacecrafts in 1975 marked the beginning of internationally coordinated space life sciences experiments performed on satellites, Soviet/Russian and US space stations and the Spacelab/Space Shuttle Systems. Since 2000 the International Space Station ISS, the biggest laboratory ever built by human mankind, is permanently inhabited in the low-Earth orbit and available for research. More than 20 years of consequent and efficient ISS utilization have delivered a multitude of new scientific findings and a wealth of fascinating results. These achievements have broadened our knowledge on basic principles of nature and have at the same time been instrumental for the development of new technologies, materials, and applications providing benefits on Earth as well as paving the way for future human and robotic exploration missions. The Chinese Space Station is another promising new platform offered by China for bilateral and international collaboration for research in life and physical sciences.

These days, space faring nations are ready to invest in the next step—exploration. In the third edition of the Global Exploration Roadmap, space agencies from around the world participating in the International Exploration Coordination Group

© Springer Nature Switzerland AG 2021
G. Ruyters et al., *Breakthroughs in Space Life Science Research*, SpringerBriefs in Space Life Sciences, https://doi.org/10.1007/978-3-030-74022-1_5

(ISECG) have formulated in 2018 a consensus about their interest in human and robotic space exploration and the importance of cooperation (https://www.globalspaceexploration.org). Further updates of the roadmap will reflect the exploration mission scenarios and efforts of space agencies; however, it does not oblige individual agencies to perform specific activities. The ISECG White Paper entitled *Scientific Opportunities Enabled by Human Exploration Beyond Low-Earth Orbit* (https://www.globalspaceexploration.org) provides a detailed description of opportunities for science and research that arise with future human-robotic exploration activities and associated infrastructures.

While the space station partners support the continuous efficient utilization of ISS in low-Earth orbit and partially have already committed their interest in its extension until 2030, NASA, ESA, and several other agencies have at the same time put forward ambitious plans for the next decade of space exploration activities in the low-Earth orbit, a gateway in cis-lunar vicinity (Lunar Orbital Platform-Gateway or Deep Space Gateway or simply Gateway), and/or on the surface of the Moon and on Mars. Exploration is expected to drive innovation and economic growth, knowledge gain, new scientific discoveries, and inspiration and not the least to provide business opportunities for the private sector. In fact, private sector entities have increasingly become interested and have started to invest in projects in and beyond low-Earth orbit under the provision of long-term governmental commitment to space exploration. Privately funded initiatives for advancing technologies and concepts have a great potential to further boost new and cost-effective science facilities, platforms, and research opportunities.

5.1   Future Priorities for Space Life Science Research

All three destinations—Gateway, Moon, and Mars—bear great potential for future space life science research. NASA's human research program focuses primarily on crew health and performance aspects, such as countermeasures and the development of capabilities and technologies to enable safe, reliable, and productive human space exploration. Based on a careful analysis of scientific results, research or knowledge gaps are identified and necessary tasks or research objectives are defined to close these gaps to mitigate risks for humans on exploration missions. Five major risks for humans traveling to Moon or Mars have been identified (https://www.nasa.gov/hrp/hazards):

- Various gravity levels ranging from microgravity during the journey, hyper-gravity conditions during launch, and hypo-gravity conditions for instance of 0.16 g on Moon and 0.3 g on Mars may lead to disorientation and balance disorders, fluid shift and visual impairment, cardiovascular deconditioning.
- Isolation/confinement may cause changes in behavior, coordination, level of commitment, and sleep disorders.

- Hostile/closed environments may lead to reduced crew performance, psychological issues, circadian desynchronization, and toxic exposure requiring a suitable spacecraft and habitation design, life support system, nutrition, and microbiome monitoring and control.
- Distance from Earth causes considerable delays in communication and requires autonomous medical care and effective emergency procedures.
- Space radiation and acute solar events, in particular, increase the long-term cancer risk and may have dramatic degeneration effects on various tissues, reproduction, and the cardiovascular and central nervous system.

ISS4Mars is an internationally coordinated initiative lead by NASA that aims to study several of the abovementioned exploration issues by using ISS as a testbed. Additional 1-year studies, operations with communication delays, human-robotic interface tests, isolation, and confinement simulations as well as sensorimotoric and performance tests after landing are foreseen to validate existing technologies and knowledge in order to better prepare for future human exploration missions. Due to long-standing partnerships between the ISS partners and various European space agencies, cooperation possibilities will certainly arise and be discussed in the frame of ISLSWG, the International Space Life Sciences Working Group.

In the past, ESA, DLR, and other European space agencies have selected their scientific research proposals predominantly based on the best-science principle rather than concentrating on crew health issues. This has ensured that important scientific findings and breakthroughs were obtained being of benefit to people on Earth. At the same time however these accomplishments were of advantage also for astronauts and space missions as such, especially in the areas of physiology, psychology, radiation, and gravitational biology. This strategy focusing on two complementary goals will certainly be continued in Europe in the exploration era.

The main objectives of space life sciences research are not expected to change significantly in the exploration era but will require some adjustments. Space biology aims to a better understanding of how different gravity levels including microgravity and other spaceflight conditions affect living biological systems, from molecule to cells, tissues, organs, and whole organisms including the human being. Exploration platforms and missions will enable biologists to collect scientific results and make discoveries in space that have enormous implications for life on Earth. Molecular and cellular repair mechanisms, muscle and bone health, metabolism, wound healing, the virulence of bacteria, viruses, and cancer, plant growth and development in life support systems and biomedical technologies are topics that are equally important for humans traveling in space, living on Moon or Mars as well as for people on Earth. Due to specific environmental conditions and changed mission requirements, the topic of bioregenerative life support systems will certainly get more attention since these systems are required—even mandatory—for complementing or finally substituting the well-known and functioning physical systems. Also, problems of long-term isolation and confinement and the associated psychological issues need to be addressed more intensely than in the era of relatively short missions with the Space Shuttle or also on stations such as MIR and the ISS.

Fig. 5.1 The Orion capsule, a lunar gateway, Moon and Mars are exploration destinations with great potential for future life sciences research. (© M. Braun, images credit ESA/NASA)

For implementing space life sciences projects, the gateway in lunar vicinity is a first and also a central part in NASA's current concept for the ARTEMIS lunar exploration program that aims to land the first woman and the next man on the Moon by 2024 and establish a sustainable human presence on the Moon in preparation for sending astronauts to Mars. When astronauts will step on the lunar South Pole, which is the first of a number of new landing sites, they will need to learn to live and work on the Moon; they need shelter from radiation, habitation and life support systems, transportation, and communication. Technologies will be developed and tested in preparation for future missions to Mars, while capabilities and crew time for research will be limited in the early phase (Fig. 5.1).

According to NASA's current ARTEMIS strategy, in a phase 1, expected to run until 2025, Gateway will only consist of a power and propulsion module, a habitation and logistics outpost to support a human moon surface expedition crew, docking ports for a cargo vehicle, and the Orion crewed vehicle as well as a human landing system. Construction of the gateway continues in phase 2 with an additional European habitation module (iHAB), the European ESPRIT module (providing refueling, infrastructure, and telecommunications), environmental control and life support systems, and more scientific instruments and research capabilities. Especially human physiology and psychology studies will benefit from continuing also research in isolation, analog, and bed rest studies on Earth.

For space, ISLSWG-organized international TIGER teams will identify human physiology, astrobiology, microbiology, and radiation biology research topics and payloads. Important European contributions to Gateway will be the internal and external dosimetry arrays consisting of active and passive dosimeters that monitor the radiation dose inside the HALO (Habitation and Logistics Outpost) module and the deep space radiation environment outside the gateway. Linear Energy Transfer (LET) spectra, total dose and accumulated dose of neutrons, electrons, protons, and ions will be recorded. These measurements will nicely complement long-term radiation measurements from inside and outside ISS.

The radiation field of the lunar vicinity is a completely new research environment outside the usually exploited low-Earth orbit conditions and is certainly one major asset of the Gateway. Its detailed impact on numerous biological processes, human physiology, biomedicine, and biotechnology as well as technologies for efficient shielding are knowledge gaps that need to be addressed and closed before humans set off to long-term deep space missions. Risks of psychological isolation and confinement, virulence of pathogens and cancer, adaptation of the microbiome, degeneration of medical products, and of delayed wound healing need to be considered. From this point of view, Gateway is of high interest for the existing science community as well as a great opportunity for new science teams.

As we explore farther from Earth, such as on the surface of the Moon or Mars, astronauts will have less access to resupply/resources from Earth. Consequently, bioregenerative and recycling components of life support, space farming, 3D bioprinting, and in situ resource utilization will become increasingly attractive with respective technology advancements becoming mandatory.

All in all, space life sciences will benefit from these new research opportunities. At the same time, space life science activities are urgently required to support long-term missions by research and development especially in the areas of bioregenerative life support systems, and for maintaining health and performance of astronauts with emphasis on psychological issues with some of the new approaches and innovative technologies described below.

5.2   New Approaches, Innovative Technologies, and Developments in Space Life Sciences for the Exploration Era

For achieving the prioritized space life sciences research goals as well as the objectives of the exploration missions as such, new approaches and additional capabilities and technologies are required. So, the greater autonomy of space crews during exploration missions requires significant progress for maintaining astronauts' health and performance. Communication delays and the absence of specific medical equipment or skills will pose new threats on mission success with medical autonomy becoming a must. Two emerging technologies that seem particularly relevant in this context are Virtual/Augmented Reality (VR and AR) and Artificial Intelligence (AI). VR and AR are increasingly applied for medical training and can also be used for guidance through complex medical procedures. Artificial Intelligence in medical decision support systems will be critical to astronauts that will need immediate assistance in exploration missions or on Moon or Mars bases.

As psychological stress will arise due to long-term isolation, working in extreme environments, being confined to narrow spaces, and other challenges, AI can act as a psychological counselor and coach the astronauts by cognitive behavioral therapy methods (Lancee et al. 2018).

Since medical consultation with the experts on ground becomes more difficult, if not impossible during exploration missions, medical platforms are not only required for monitoring of health and performance of astronauts and for diagnosing and providing feedback, but also for training and for continuous learning. Integrated Intelligent Health Evaluation Platforms such as described by Nasseri and Impersario (2018) combine various technologies and devices thus enabling as much autonomy as possible for the astronauts.

Another aspect relates to personalized medicine supported by "omics." In the past, medical treatments and countermeasures for astronauts in space were elaborated, tested, and applied for the "typical" human or astronaut. A better understanding of individual differences may allow the development of individualized countermeasures, which optimize the safety and performance of each astronaut entering the space environment. In this context, it has to be considered that the morphological, physiological, and behavioral phenotypes are governed by an underlying molecular network of vast complexity bringing "omics" into play (Schmidt et al. 2017). As lessons learned from the twin study on ISS, NASA is attempting slowly to shift from the previous "one-size-fits-all" approach to a more personalized care for individuals, sometimes called "precision medicine, by omics" (www.nasa. gov; press release Aug 4, 2016; see Sect. 3.5.7). Omics integrates various biological disciplines to focus on measurements of a diverse array of biomolecules. It combines genomics, transcriptomics, proteomics, epigenomics, metabolomics, and microbiomics to arrive at a larger picture of the human body at a fundamental level. Before this approach can be applied to its full picture, the challenge of storing, interpreting, and securing huge amounts of data needs to be met. IBM's Medical WATSON may be a promising step in this direction (see Sect. 5.2.2).

In the following, specific examples for devices and technologies required in the exploration era are described that are currently under consideration or under development or even in the testing phase—worldwide, but with some emphasis on European activities. To close important technological gaps, items already available on ground but requiring upgrades or modification for space utilization are also considered.

5.2.1 Medical Imaging Technologies for Human Exploration

After summarizing technologies and devices that have been developed for space in the area of (non-invasive) medical diagnostics in Chap. 4, examples for technologies required will be discussed first that are in use on Earth but have not yet found their way for space application.

A few remarks on medical imaging technologies in general and about their space application in particular are made before describing pQCT as a method for bone and muscle imaging. Medical Imaging is defined as a technique of creating visual representations of the interior of a body for physiological and clinical analysis. In contrast, measurement and recording techniques such as EEG and ECG produce data rather than images.

Regarding ultrasound imaging, CNES has shown some activities in the past. Firstly, CNES developed in 1982 a Doppler scanner device called "As de Coeur" for their MIR missions between 1988 and 1995 (Seibert 2001). Later ECHO was developed that is a tele-operated ultrasound scanner that allows a specialist to examine the astronauts almost autonomously from the ground. Quite recently, namely in September 2016, an ultrasound device was launched in cooperation with China in the frame of the Cardiospace project onboard of the new TianGong-2 orbital module (www.terason.com). NASA aims to advance from 2D to 3D within the next few years and has recently used ultrasound on the ISS for monitoring changes in the lumbar and sacral spine (for an overview about ultrasound application in space, see www.hagerstownccc.edu).

For investigations on bone loss in space, a bone scanner had been developed by TechShot using dual-energy X-ray absorptiometry (DXA). Supported by funding from NASA and CASIS (Center for the Advancement of Science in Space), a DXA device was qualified for use on the ISS in 2015. DXA is widely used in the study of bone loss in laboratory animals, although it does not provide information about bone microstructure, which is mandatory to assess bone strength. According to TechShot and NASA reports, the space-based DXA device has successfully displayed bone mineral density, bone mineral content, lean mass, fat mass, total mass, and percent fat; however, scientific results have not yet been published to our knowledge.

5.2.1.1 HR-pQCT

High-resolution peripheral quantitative computed tomography (HR-pQCT), well established in everyday clinical practice for assessing the bone microstructure, quality and strength in vivo, has to our knowledge not been developed for utilization in space. This is why bone analysis via pQCT on astronauts is still being performed only pre- and postflight. For long-term exploratory missions, this seems not to be acceptable so that an important technological gap exists here.

pQCT provides an automatic scan analysis of trabecular and cortical bone compartments, calculating not only their bone mineral density but also images the bone geometrical parameters such as marrow and cortical cross-sectional area, cortical thickness, both periosteal and endosteal circumference, as well as biomechanical parameters like cross-sectional moment of inertia, a measure of bending, polar moment of inertia, indicating bone strength in torsion, and strength strain index. Also, the cross-sectional area of muscle and fat can be extracted. Thus, pQCT provides an evaluation of the functional muscle-bone unit (Stagi et al. 2016).

Recently HR-pQCT (Three-dimensional high-resolution quantitative computed tomography) has introduced a new dimension in the imaging of bone and joints by providing images in vivo that are in three dimensions and in high resolution while at the same time exposing the human subject to very low levels of radiation. For pre- and postflight measurements on ISS astronauts, but also in various bed rest studies organized by ESA, CNES, and DLR, the so-called Xtreme CT by the Swiss company Scanco Medical was used to assess density and microarchitecture of the bone tissue (Belavý et al. 2011).

According to Scanco Medical, the XtremeCT II is the most recent new generation high-resolution peripheral quantitative computed tomography (HR-pQCT). The XtremeCT II is designed to measure the bone density and to quantify the three-dimensional microarchitecture of the bone at the distal tibia and radius of humans for clinical in vivo assessment of osteoporosis at an even higher precision and speed than its predecessor, the XtremeCT (www.scanco.ch). A low X-ray dose allows regular follow-up measurements, and powerful true-3D evaluation software automatically matches cortical and trabecular bone regions to previous measurements for direct comparison of density and structure characteristics.

In Germany, the company Novotec Medical/STRATEC Medizintechnik (www.galileo-training.com) provides pQCT devices that are widely used for scientific and diagnostic purposes and were applied also in recent bed rest studies. Interestingly, the device is certified for operation without applying X-ray shielding of the subject. It remains to be seen if this technology will be made available also for use in space in the future.

5.2.1.2   XRMON and XRF: Radiography and Radioscopy on ISS

X-ray radiography and radioscopy investigations onboard the International Space Station are primarily intended for in situ materials processing research but bear also a great potential for biological investigations. For almost two decades DLR and ESA have funded an X-ray multiuser diagnostics facility (XRMON developed by Airbus Defence and Space, Germany) that has already been flown on sounding rockets (MASER, MAXUS, and MAPHEUS) and parabolic flights of airplanes to study metal-alloy solidification, metal foaming, and diffusion in metallic melts as well as granular matter. Based on vast experience gained from these hardware and experiment heritage and driven by the scientific need for X-ray diagnostics in a stable microgravity environment, a compact X-ray radiography facility is planned as ISS payload.

A compact X-ray microtomography device that had been built as a technology demonstrator in a study supported by ESA had already shown the suitability of a low-powered microfocus X-ray tube for element mapping and absorption imaging by exploiting scanning fluorescence tomography and by full-field transmission microtomography (Feldkamp et al. 2007). An X-ray device for the ISS containing a small high-energy X-ray source would support fluorescence microtomography when employing a rotation stage. Potential biological investigations could include on-site monitoring of changes in three-dimensional structures like bones of small animals, microbial growth on a substrate as well as mineralization in an organoid culture. Also, two-dimensional fluorescence element mapping would facilitate the analysis of changes of the element composition in biological material, monitoring of minerals and toxic elements in water like Pb, Cd, Hg, nutrients, and toxic elements in bones, organs, and plants. In addition, acute radiation doses could be applied in combination with microgravity conditions to properly simulate conditions of a deep space environment during long-term traveling or living in space. In particular, the visualization of metabolites within biological tissues and organs promises new insights into the effects of radiation exposure and altered gravity conditions on physiological pathways and biological processes that eventually lead to the adverse effects of space conditions on living organisms.

5.2.2 Autonomous Artificial Intelligence Assistance

Development of human-machine interfaces that support astronauts, relieve social monotony, and engender loyalty and trust are recommended projects in many space agencies as expressed, e.g., in the NASA Technology Roadmap of 2015 or in the European THESEUS strategy document of 2012 (https://www.nasa.gov/sites/default/files/atoms/files/2015_nasa_technology_roadmaps_ta_6_human_health_life_support_habitation_final.pdf; http://archives.esf.org/fileadmin/Public_documents/Publications/RoadMap_web_01.pdf).

Several space agencies have made efforts to develop more or less autonomous artificial intelligence (AI)-based assistance systems for the ISS. The capabilities of these devices are rapidly increasing. While JAXA's INT-Ball currently focuses on the tasks of a drone (photo and video documentation) that later shall be able to autonomously cruise the station, NASA's Astrobee is a free-flying robotic system which is supposed to help with everyday routine tasks like inventorying, documenting experiments (for details see https://www.nasa.gov/astrobee), and experiment conduction. DLR entered the stage of intelligent astronaut support with CIMON, the Crew Interactive Mobile Companion. CIMON is a mobile and autonomous assistance system for astronauts that was developed by Airbus, based on IBM's artificial intelligence platform WATSON. The technology-demonstration project was supported by scientists from the LMU Munich on contract by the German DLR Space Agency. A first test was performed on ISS by the German ESA astronaut Alexander Gerst in 2018, and also Luca Parmitano enjoyed communicating with CIMON in 2020. The medicine-ball-sized free-flying CIMON demonstrator weighs about 5 kg; its basic structure was manufactured by 3D printing (Eisenberg et al. 2018) (Fig. 5.2).

CIMON was designed to verbally interact with the astronauts, navigate by verbal commands, recognize voices and faces, understand commands, communicate with and support astronauts in performing routine work, for example, by reading instructions or—thanks to its "neural" AI network and its ability to learn—offering

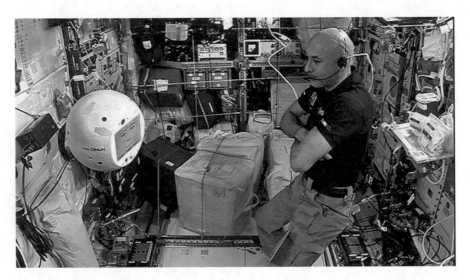

Fig. 5.2 The Italian ESA astronaut Luca Parmitano communicates with the free-floating artificial intelligence companion CIMON testing "his" navigation and communication skills. (© ESA/NASA)

solutions for problems. CIMON uses Watson AI technology from the IBM cloud and, with its digital face, voice, and artificial intelligence, has the potential to become a helpful "colleague" onboard as well as an empathic companion. With CIMON, crew members can do more than just work through checklists and procedures; they can also engage with their AI assistant. In this way, CIMON makes work easier for the astronauts when carrying out everyday routine tasks, helps to increase efficiency and mission success, and improves security, as it can also serve as an early warning system for technical problems.

CIMON was trained to recognize its environment and its human interaction partners. CIMON's AI has text, speech, and image processing capabilities as well as the ability to retrieve specific information and to collect data. These skills, which can be trained individually and deepened in the context of a given assignment, are developed based on the principle of understanding—reasoning—learning.

In its first mission on ISS in 2018, CIMON-1 was equipped with a selected range of capabilities. Alexander Gerst tested the basic functions of CIMON and successfully demonstrated CIMON's orientation and active positioning in weightlessness, reaction to voices, and receive and obey orders. Two years later, Luca Parmitano tested CIMON-2, an updated version of the demonstrator, for its autonomous flight capabilities, voice control system, and taking images and videos.

In the medium term, aerospace researchers also plan to use the CIMON project to examine group effects that can develop over a long period of time in small teams and that may arise during long-term missions to the Moon or Mars. Social interaction between people and machines, between astronauts and assistance systems, equipped with emotional intelligence, could play an important role in the success of long-term exploration missions (Eisenberg et al. 2018). Obviously, also Earth applications are manifold, e.g., several initiatives have already been started to explore the possible uses of such intelligent systems in hospitals, health care, and social systems.

5.2.3   CRISPR/CAS

The CRISPR/CAS method (Clustered Regularly Interspaced Short Palindromic Repeats) is a biochemical technology to cut and modify DNA. The first paper about its development and application was published in 2012 by Charpentier and Doudna and their coworkers (Jinek et al. 2012). In 2015, Science declared the CRISPR method the "Breakthrough of the Year 2015" (www.sciencemag.org). In 2020, the Nobel Prize in Chemistry was awarded to Emmanuelle Charpentier and Jennifer Doudna for discovering the CRISPR/Cas9 genetic scissors. Researchers can use these to change the DNA of animals, plants, and microorganisms with extremely high precision; genes can be inserted, deleted, or shut off; and even nucleotides can be modified. This technology has revolutionized the molecular life sciences, brought new opportunities for plant breeding, is contributing to innovative cancer therapies, and may make the dream of curing inherited diseases come true.

CRISPR can be used to target and modify DNA with groundbreaking accuracy. While several gene editing techniques have had success in the last decade, CRISPR's ease of use and affordability allows nearly any molecular biology research lab and even some high schools to make use of it. The research it has helped advance include disabling retroviruses encoded in the pig genome that have long posed a safety concern for transplanting organs from pigs to humans; developing a gene drive, which allows an introduced gene to be transmitted throughout pest populations faster than naturally possible, offering hope for combatting vector-borne illnesses; and the first deliberate editing of the DNA of human embryos. While this last revolutionary development leads scientists and doctors to hope that potential genetic therapies and treatments will soon be discovered for various diseases, it also raises concerns about the necessity of having an international summit on human genetic engineering. In Germany, the scientific society Leopoldina and Spektrum have published specific contributions on this topic (www.leopoldina.org; Ledford 2015).

Nevertheless, as *Science* Editor-in-Chief Marcia McNutt notes, "Researchers have long sought better ways to edit the genetic code in cultured cells and laboratory organisms to silence, activate, or change targeted genes to gain a better understanding of their roles. This, in turn, could open the door to beneficial applications, from ecological to agricultural to biomedical." CRISPR may well be the tool that researchers have been waiting for to bring theory to practical usage (for review see www.wikipedia.org, including references cited).

As far as we know, the first experiment in space has been performed on the ISS in 2019 in the frame of the "Genes in Space Program" (www.issnationallab.org). In trying to develop an opinion on the importance of this technique in context with future exploration missions, NASA has recently published a study authored by

Hendrickson (2016) with the title "Is there a use for CRISPR/CAS9 in space?" (www.three.jsc.nasa.gov/articles/CRISPR.pdf). Considering all these aspects, it certainly seems worthwhile to initiate discussions on the possibilities and chances but also on ethical considerations of using CRISPR/CAS technology for human space exploration within international settings such as the International Space Life Sciences Working Group (ISLSWG).

5.2.4 3D Cell Cultures, Organoids-on-a-Chip, and 3D-Bioprinting

Biomedical investigations in space are highly dependent on astronauts and cosmonauts as human test subjects. Alternatives using animal models are also extraordinarily complex, subject to ethical constraints, and of technical and practical limitations. Many aspects such as immune system dysregulation, altered osteoclast and osteoblast function, and many more have been intensively studied by means of monolayer cell culture experiments which are much more easily implemented on ISS but come with the disadvantage of limited validity and transferability of findings with respect to effects on the whole organism and the organ-specific physiological processes involved.

3D-Cell culture systems increasingly receive the attention of space life sciences researchers for multiple reasons; most importantly, cell aggregates resemble more closely in structure and function the composition of organs and tissues while providing many advantages of the in vitro culture systems. More recently, organoids-on-a-chip designs have evolved quickly that make use of microfabricated cell culture devices with microfluidic and even biophysical manipulation capabilities which are able to mimic physiological responses of human organ units more accurately than 2D cultures or animal models (Skardal et al. 2016). 3D-cell culture models may either be formed by self-organizing stem cell aggregates or created by additive-layered manufacturing. Such a 3D-bioprinting process can involve printing of cells in a 3D matrix or a nondegradable scaffold thus imitating, e.g., functional bone or bone marrow units.

Although the 3D-bioprinting technology still has a long way to go until cell aggregates are successfully matured into fully functional tissues or larger organs on Earth, tissue-specific constructs promise to dramatically improve our understanding of the biophysical mechanisms of tissue generation, regeneration, and physiological responses to environmental cues like gravity, altered gravity, and radiation. The potential of the 3D-bioprinting technology is obvious and includes fundamental research in materials and processes, in tissue/organoid engineering, biotechnology, and synthetic biology, in biological processes on cell, tissue, and organoid level in response to changing environmental parameters like gravity and radiation. In a more technical sense, 3D-bioprinting technologies might be used to manufacture microalgae-based constructs for the production of medication, nutritional supplements, and life support components, but also medical tools, tailored casts, and dental equipment such as dental implants and fillings. Ultimately, 3D-bioprinting might be used to produce a wide range of implantable patient-specific tissue constructs for treating skin lesions and bone defects or even patient-specific vascularized tissue constructs like kidney or liver (Cubo-Mateo et al. 2020). Bioprinting methods promise great progress in regenerative and personalized medicine for astronauts as well as for people on Earth, for radiation investigations on specific tissues and organs, and also pharmacological and pharmaceutical purposes. In a long-term perspective, 3D-bioprinting and additive-layered manufacturing may provide the fundamental technology for autonomous medical treatment and for regenerative life support in habitats on Moon or Mars and on manned space exploration missions.

Space agencies from around the world have started activities to exploit the potentials of 3D-bioprinting technologies in space. Especially NASA is highly active in testing new technologies. In 2016, NASA brought the MinION DNA sequencer onboard of the ISS. For the first time, a simulated real-time analyses of in-flight sequence data was performed and demonstrated the feasibility of sequencing analysis and microbial identification aboard the ISS. These findings illustrate the potential for sequencing applications including disease diagnosis, environmental monitoring, and elucidating the molecular basis for how organisms respond to spaceflight (Castro-Wallace et al. 2017). In 2017, CASIS (Center for the Advancement of Science in Space) together with NCATS (National Center for Advancing Translational Sciences) as part of the National Institutes of Health, (NIH) selected five new projects for ISS research, funded by NASA, featuring tissue chips for improving human health and disease research via translational research. Interestingly, also a project under the leadership of German scientists focusing on immunological senescence and its impact on tissue stem cells and regeneration is among the new projects (https://ncats.nih.gov/tissuechip/projects/space2017).

All in all, these upcoming new technologies for ISS research during long-duration exploration missions will certainly also fertilize respective developments for the benefit of people on Earth.

5.3 Final Thoughts: From World War II to Long-Term Exploration of Space—70 Years of Space Life Science Research

5.3.1 A View in Retrospect

More than 70 years after introducing the term "space medicine" and the foundation of the world's first department of space medicine at the Randolph Airforce Base, 60 years after the first human spaceflights of Yuri Gagarin and John Glenn in 1961 and 1962, respectively, and 20 years after the first astronauts have set their feet into the ISS, space life sciences have achieved worldwide recognition as important complementation of science and development. In the first years or even decades of (human) spaceflight, the focus of space life sciences activities certainly was laid on survival and maintaining health and performance of astronauts thereby ensuring the success of the missions. However, scientists recognized soon the chances to perform also fundamental or basic research under the new and extraordinary conditions provided by spaceflight, especially by making use of microgravity. For the first time ever, it became possible to shut off the constantly acting force of gravity for elucidating its role for evolution, distribution, and functioning of life acting on Earth for about 4 billion years.

Operational constraints, often making experimentation in space tedious and difficult, at the same time stimulated (international) cooperation of scientists and space agencies and led medical doctors to view the human body as an integrated system—a view that is shared more and more by their colleagues on Earth after decades of looking at the various physiological system of humans separately, i.e., as experts just for bone, heart, brain so-to-say. Space life sciences have evolved over the decades from observational studies that looked at the effects of microgravity on cells, microorganisms, plants, animals, and humans to the thorough analysis of mechanisms on the cellular and molecular level as is the case in the terrestrial lab. It took also quite a while for the scientific community to understand and accept that space life sciences do not represent a new, separate research field, but that space life sciences projects are meant to continue and complement terrestrial research topics under special conditions with novel research opportunities.

In addition to various infrastructure elements developed over time for human spaceflight such as the Space Shuttle system and space stations from Skylab via Salyut, MIR to the ISS, additional flight opportunities were established. They provide microgravity conditions from a few seconds such as the Bremen Drop Tower and parabolic airplane flights, minutes as in sounding rockets like TEXUS up to days or weeks in satellites. Also, possibilities for simulating space conditions, especially microgravity, on Earth have contributed significantly to the success of life sciences research. They range from clinostats, centrifuges, and random positioning machines for biological research up to complex isolation, confinement, and bed rest studies with human subjects for performing physiological and psychological investigations. It is fair to say that—without the space life sciences programs of the various space agencies—these complex and expensive international bed rest campaigns and long-term isolation studies such as Mars500 with their wealth of scientific accomplishment would never have materialized.

By these studies in space and on Earth, space life sciences research has led to convincing stepwise progress in many areas of biology, biotechnology, and medicine and to breakthroughs and new concepts for some topics. Among them, we find the (nearly) complete elucidation of the signal transduction chain for gravitaxis and gravitropism in microorganisms and plants as well as the completely new integrative view on the context of salt with nutrition, the bone and immune system, and cardiovascular problems including hypertension leading to novel diagnosis and treatment options. Thus, from the molecular and cellular level up to the human being, space life sciences have proven their benefit for people on Earth. In addition, the development of special innovative devices for monitoring, diagnostics, and treatment of diseases—often non-invasive, robust, reliable, small, and easy to handle—has accompanied and even enabled these scientific accomplishments with significant application potential on Earth, in the clinic or in elderly people at home. The same is true for the development of countermeasures for rehabilitation and maintaining fitness on Earth that has been stimulated by spaceflight also for the benefit of the aging society.

5.3.2   A Look into the Future

The challenges confronting health care especially in today's aging societies of the industrialized countries demand a paradigm change in the medical sector. Among experts, it is generally expected that medicine and health care will focus on three tasks in the future:

- Health maintenance in healthy people.
- Individualized health care.
- Challenging the traditional relation between physician and patient by providing the doctor's advice irrespective of the location of the patients (telecare).

Now, what has all that to do with space medicine? The paradigm changes postulated have basically been realized in space life sciences and medicine since many years. Medical research and health care in space have always been and still are focusing, in an integrative approach, on the healthy astronaut as an individual, and telemedicine is being practiced for decades as well (for review see Ruyters and Stang 2016). From this, terrestrial medicine, and healthcare stand to learn much from space life sciences in the future.

Medical spaceflight experiments with their integrated view on the human body have also guided the scientific experts to early detect the similarity of the physiological and psychological changes experienced by astronauts during their spaceflights with those of the aging process on Earth. In fact, the changes in space occur in time-lapse so that the results of the scientific projects including testing the efficacy of countermeasures can be achieved much faster as during the aging process. Different from Earth, (most of) the changes in astronauts are reversible so that also the process of readaptation to Earth's gravity can be studied leading to further interesting results on the adaptability of the physiological systems of humans. Since health problems associated with aging in the (Western) industrialized population are judged as one of the greatest challenges for the time being and even more so in the future, the contribution of space life sciences in solving or at least in understanding these problems cannot be overestimated.

From these considerations, it is obvious that—especially for people living in an aging but nevertheless active society—space medicine is of inestimable importance, witnessed by the numerous scientific findings and innovative non-invasive diagnostic devices developed in the recent past. In the years to come, space medicine and space life sciences are likely to become even more powerful drivers for change in terrestrial medicine.

Now is the time to take the next steps to go beyond low-Earth orbit and explore Moon, Mars, and other distant destinations. Space Life Science has to and will contribute to achieving these new goals. New challenges, chances, and opportunities of the long-duration exploration missions will further stimulate space life science research, which needs to be performed in a highly autonomous manner thus continuing and multiplying the success stories of space life science research and development described in the present book.

References

Belavý DL, Beller G, Ritter Z, Felsenberg D (2011) Bone structure and density via HR-pQCT in 60d bed-rest, 2-years recovery with and without countermeasures. J Musculoskelet Neuronal Interact 11:215–226

Castro-Wallace SL, Chiu CY, John KK, Stahl SE, Rubins KH, McIntyre ABR, Dworkin JP, Lupisella ML, Smith DJ, Botkin DJ, Stephenson TA, Juul S, Turner DJ, Izquierdo F, Federman S, Stryke D, Somasekar S, Alexander N, Yu G, Mason CE, Burton AS (2017) Nanopore DNA sequencing and genome assembly on the International Space Station. Sci Rep 7:18022. https://doi.org/10.1038/s41598-017-18364-0

Cubo-Mateo N, Podhajsky S, Knickmann D, Slenzka K, Ghidini T, Gelinsky M (2020) Can 3D bioprinting be a key for exploratory missions and human settlements on the Moon and Mars? Biofabrication 12:043001

Eisenberg T, Schulien P, Kössl C, Buchheim JI, Biniok M, Karrasch C, Schmid V (2018) CIMON – a mobile artificial crew mate for the ISS. 69th IAC Bremen. IAC-18,B3,7,8x46149

Feldkamp JM, Schroer CG, Patommel J, Lengeler B, Günzler TF, Schweitzer M, Stenzel C, Dieckmann M, Schroeder WH (2007) Compact x-ray tomography system for element mapping and absorption imaging. Rev Sci Instrum 78:073702. https://doi.org/10.1063/1.2751094

Hendrickson EA (2016) Precise genome engineering and the CRISPR revolution (boldly going where no technology has gone before). https://three.jsc.nasa.gov/articles/CRISPR.pdf

Jinek M, Chylinski K, Fonfara I, Hauer M, Doudna JA, Charpentier E (2012) A programmable dual-RNA–guided DNA endonuclease in adaptive bacterial immunity. Science 337:816–821

Lancee GJ, Engelen LJ, van de Belt TH (2018) Medical autonomy as prerequisite for deep space travel will benefit from terrestrial healthcare innovation. 69th IAC Bremen. IAC-18. A5.2.7x47898

Ledford H (2015) Werkzeug der Genmanipulation – Gentechnik: CRISPR verändert alles. https://www.spektrum.de/news/gentechnik-crispr-erleichtert-die-manipulation/1351915

Nasseri SA, Impersario G (2018) The concept of an integrated intelligent health evaluation and support platform for deep space exploration. 69th IAC Bremen. IAC-18-B.3.9-GTS.2

Ruyters G, Stang K (2016) Space medicine 2025 – a vision. REACH Rev Human Space Explor 1:55–62

Schmidt MA, Goodwin TJ, Cattino M (2017) Personalized medicine in space flight. In: Legato MJ (ed) Principles of gender-specific medicine. Academic, Cambridge, MA. Part 1, pp 655–672, and Part 2, pp 673–693

Seibert G (2001) A world without gravity – research in space for health and industrial processes. ESA SP-1251

Skardal A, Shupe T, Atala A (2016) Organoid-on-a-chip and body-on-a-chip systems for drug screening and disease modeling. Drug Discov Today 21(9):1399–1411. https://doi.org/10.1016/j.drudis.2016.07.003

Stagi S, Cavalli L, Cavalli T, de Martino M, Brandi ML (2016) Peripheral quantitative computer tomography (pQCT) for the assessment of bone strength in most of bone affecting conditions in developmental age: a review. Ital J Pediatr 42. https://doi.org/10.1186/s13052-016-0297-9

Printed in the United States
by Baker & Taylor Publisher Services